GRAFT COPOLYMERS OF POLYSACCHARIDES FOR DRUG DELIVERY SYSTEMS

GRAFT COPOLYMERS OF POLYSACCHARIDES FOR DRUG DELIVERY SYSTEMS

Rishabha Malviya, PhD
Sonali Sundram, MPharm
Editors

AAP | APPLE ACADEMIC PRESS

First edition published 2025

Apple Academic Press Inc.
1265 Goldenrod Circle, NE,
Palm Bay, FL 32905 USA

760 Laurentian Drive, Unit 19,
Burlington, ON L7N 0A4, Canada

CRC Press
2385 NW Executive Center Drive,
Suite 320, Boca Raton FL 33431

4 Park Square, Milton Park,
Abingdon, Oxon, OX14 4RN UK

Library and Archives Canada Cataloguing in Publication

Title: Graft copolymers of polysaccharides for drug delivery systems / Rishabha Malviya, PhD, Sonali Sundram, PhD, editors.
Names: Malviya, Rishabha, editor. | Sundram, Sonali, editor.
Description: First edition. | Includes bibliographical references and index.
Identifiers: Canadiana (print) 2025012792X | Canadiana (ebook) 20250127938 | ISBN 9781774919323 (hardcover) | ISBN 9781774919330 (softcover) | ISBN 9781003597995 (ebook)
Subjects: LCSH: Drug delivery systems. | LCSH: Graft copolymers. | LCSH: Polysaccharides—Synthesis.
Classification: LCC RS199.5 .G73 2025 | DDC 615/.6—dc23

Library of Congress Cataloging-in-Publication Data

CIP data on file with US Library of Congress

ISBN: 978-1-77491-932-3 (hbk)
ISBN: 978-1-77491-933-0 (pbk)
ISBN: 978-1-00359-799-5 (ebk)

About the Editors

Rishabha Malviya, PhD
Associate Professor, Department of Pharmacy, School of Medical and Allied Sciences, Galgotias University, India

Rishabha Malviya, PhD, has been working as an Associate Professor in the Department of Pharmacy, School of Medical and Allied Sciences, Galgotias University for the past 8 years. He completed a B. Pharmacy from Uttar Pradesh Technical University and an M. Pharmacy (Pharmaceutics) from Gautam Buddha Technical University, Lucknow, Uttar Pradesh. His PhD (Pharmacy) work was in the area of novel formulation development techniques. He has 12 years of research experience. His areas of interest include formulation optimization, nanoformulation, targeted drug delivery, localized drug delivery, and characterization of natural polymers as pharmaceutical excipients. He has authored more than 200 research/review papers for national/international journals of repute. He has 58 patents (19 granted, 38 published, 1 filed) and publications in reputed national and international journals with a total of more than 200 cumulative impact factors. He has also received an Outstanding Reviewer award from Elsevier. He has authored/edited 48 books (Wiley, CRC Press/Taylor and Francis, Springer, River Publisher, IOP Publishing, and OMICS Publishing) and authored 31 book chapters. His name has been included in the world's top 2% scientist list for the years 2020 and 2021 by Elsevier BV and Stanford University. He is a reviewer/editor/editorial board member of more than 50 national and international journals of repute. He has been invited as an author for "Atlas of Science" and Pharma magazine dealing with industry (B2B) "Ingredient South Asia Magazines."

Sonali Sundram, MPharm
Assistant Professor, School of Medical and Allied Sciences, Galgotias University, Greater Noida, Uttar Pradesh, India

Sonali Sundram, MPharm, is currently working at Galgotias University, Greater Noida. She completed a B. Pharm and M. Pharm (Pharmacology) from AKTU, Lucknow. She has worked as a research scientist on an ICMR project at King George's Medical University, Lucknow; after that, she joined BBDNIIT. Her PhD (Pharmacy) work was in the area of neurodegeneration and nanoformulation. Her areas of interest are neurodegeneration, clinical research, and artificial intelligence. She has authored/edited more than 15 books (Wiley, CRC Press/Taylor and Francis, IOP Publishing, Apple Academic Press/Taylor and Francis, Springer Nature, and River Publisher). She has attended and organized more than 15 national and international seminars/conferences/workshops. She has more than 8 national and international patents to her credit. She has published six SCI-indexed manuscripts (cumulative impact factor: 20.71) with reputed international publishers. She has delivered oral presentations at international conferences organized in different European countries.

Contents

Contributors

Chembuli Abhishek
Master Student, Department of Biotechnology, North-Eastern University, USA

Michael U. Adikwu
Department of Pharmaceutics, University of Nigeria, Nsukka, Nigeria

Usama Ahmad
Associate Professor, Faculty of Pharmacy, Integral University, Lucknow, Uttar Pradesh, India

Isaac O. Babatunde
Phytomedicine and Molecular Toxicology Research Laboratories, Department of Biochemistry, Osun State University, Osogbo, Nigeria

Ajaykumar Chittipolu
Assistant Professor, Vision College of Pharmaceutical Sciences and Research, JNTUH, Hyderabad, Telangana, India

Syed Fareha
Department of Bioengineering, Integral University, Lucknow, Uttar Pradesh, India

Anas Islam
Faculty of Pharmacy, Integral University, Lucknow, Uttar Pradesh, India

Mohd. Muazzam Khan
Faculty of Pharmacy, Integral University, Lucknow, Uttar Pradesh, India

H. C. Ananda Murthy
Department of Applied Chemistry, School of Natural Science, Adama Science and Technology University, Adama, Ethiopia

Bankole E. Ofeniforo
Department of Biochemistry, Faculty of Life Sciences, University of Ilorin, Ilorin, Kwara, Nigeria

Johnson O. Oladele
Biochemistry Unit, Department of Chemical Sciences, Faculty of Sciences, Kings University, Ode-Omu, Osun, Nigeria; Phytochemistry and Phytomedicine Research Unit, Royal Scientific Research Institute, Osun, Nigeria

Oluwaseun T. Oladele
Phytomedicine and Molecular Toxicology Research Laboratories, Department of Biochemistry, Osun State University, Osogbo, Nigeria

Emmanuel O. Olorunsola
Department of Pharmaceutics and Pharmaceutical Technology, University of Uyo, Uyo, Nigeria

Aristeidis Papagiannopoulos
Theoretical and Physical Chemistry Institute, National Hellenic Research Foundation, Athens, Greece

Akbar Pasha
Research Scholar, Department of Genetics and Biotechnology, Osmania University, Hyderabad, Telangana, India

Natassa Pippa
Department of Pharmaceutical Technology, Faculty of Pharmacy, Panepistimioupolis Zografou, National and Kapodistrian University of Athens, Athens, Greece

Stergios Pispas
Theoretical and Physical Chemistry Institute, National Hellenic Research Foundation, Athens, Greece

Bhupendra Prajapati
Shree S. K. Patel College of Pharmaceutical Education and Research, Ganpat University, Mehsana, Gujarat, India

Doneti Ravindar
Research Scholar, Department of Genetics and Biotechnology, Osmania University, Hyderabad, Telangana, India

Nensi Raytthatha
Faculty of Pharmacy, Sigma University, Vadodara, Gujarat, India

S. Giridhar Reddy
Department of Chemistry, Amrita School of Engineering, Bangalore, Amrita Vishwa Vidyapeetham, India

Theodore Sentoukas
Center of Polymer and Carbon Materials, Polish Academy of Sciences, Zabrze, Poland

Efstathia Triantafyllopoulou
Department of Pharmaceutical Technology, Faculty of Pharmacy, Panepistimioupolis Zografou, National and Kapodistrian University of Athens, Athens, Greece

Jigar Vyas
Faculty of Pharmacy, Sigma University, Vadodara, Gujarat, India

Abbreviations

AAc	Acrylic Acid
AFM	Atomic Force Microscopy
AIBN	Azobisisobutyronitrile
AN	Acrylonitrile
APIs	Active Pharmaceutical Ingredients
BCS	Biopharmaceutics Classification System
BEV	Bevacizumab
CA	Citraconyl Anhydride
CAD	Computer-Aided Design
CDA	Cellulose Diacetate
CG	Cashew Gum
CMC	Carboxymethyl Cellulose
CMC	Critical Micelle Concentration
CRP	Controlled/Living Polymerization
CRS	Controlled Release System
CRTs	Controlled-Release Technologies
CS	Chitosan
CTA	Chain Transfer Agent
DASNPs	Dialdehyde Starch Nanoparticles
DDA	Dodecylamine
DDS	Drug Delivery System
DEX	Dexamethasone
DLS	Dynamic Light Scattering
DOX	Doxorubicin
DSC	Differential Scanning Calorimetry
EGDMA	Ethylene Glycol Dimethacrylate
ELS	Electrophoretic Light Scattering
EPR	Enhanced Permeability and Retention
FTIR	Fourier Transform Infrared Spectroscopy
GAGs	Glycosaminoglycans
GFN	Guaifenesin
GG	Guar Gum
GGS	Guar Gum Succinate
GI	Gastrointestinal
GPC	Gel Permeation Chromatography
GSH	Glutathione
HA	Hyaluronic Acid

HAS	Human Serum Albumin
HER2	Anti-Human Epidermal Growth Factor Receptor 2
HFV	Human Foamy Virus
hMSCs	Human Mesenchymal Stem Cells
HPC	Hydroxypropyl Cellulose
HPMC	Hydroxypropyl Methylcellulose
ICT	Icaritin
IDDS	Innovative Drug Delivery System
IONs	Iron Oxide Nanoparticles
IPN	Interpenetrating Polymer Network
IR	Infrared
KFs	Keratin-Based Formulations
LCST	Lower Critical Solution Temperature
LMW	Low Molecular Weight
MA	Methacrylamide
MA	Methacrylic Acid
mAbs	Monoclonal Antibodies
MC	Methylcellulose
MEMS	Microelectromechanical Systems
MMT	Montmorillonite
MNP	Magnetite Nanoparticles
MNs	Microneedle
MSNPs	Mesoporous Silica Nanoparticles
NC	Nanocellulose
NIPAM	N-Isopropyl Acrylamide
NMR	Nuclear Magnetic Resonance
NVP	N-Vinylpyrrolidone
OEI	Oligo Ethylenimine
OFET	Organic Field-Effect Transistors
PA	Peritendinous Adhesion
PAA	Poly (Acrylic Acid)
PAM	Polyacrylamide
PCGA	Poly(ε-Caprolactone-Co-Glycolide)
PCL	Poly (ε-caprolactone)
PCLA	Poly(ε-Caprolactone-Co-Lactide)
pDNA	Plasmid DNA
PEA	Poly (Ester Amide)
PEG	Poly (Ethylene Glycol)
PEI	Polyethyleneimine
PEO	Poly (Ethylene Oxide)
PetOx	Poly(2-Ethyl-2-Oxazoline)
PGA	Polyglycolide
PIL	Poly (Ionic Liquid)

PKs	Pharmacokinetics
PLA	Polylactic Acid
PLGA	Poly (Lactic-Co-Glycolic Acid)
PLH	Poly(L-Histidine)
PMMA	Poly (Methyl Methacrylate)
PNIPAM	Poly(N-Isopropylacrylamide)
PP	Polypropylene
PPO	Poly (Propylene Oxide)
PPy	Polypyrrole
PPZs	Polyphosphazene
PS	Polystyrene
PSS	Polystyrene Sulfonate
PTX	Paclitaxel
PVA	Polyvinyl Alcohol
PVCL	Poly (N-Vinylcaprolactam)
QCD	Quaternized Cyclodextrin
RAFT	Reversible Addition-Fragmentation Chain Transfer
RC	Rubber Chain
RES	Reticuloendothelial System
rGO	Reduced Graphene Oxide
ROP	Ring Opening Polymerization
SANS	Small Angle Neutron Scattering
SAS	Small Angle Scattering
SAXS	Small Angle X-Ray Scattering
SEDDS	Self-Emulsifying Drug Delivery Systems
SEM	Scanning Electron Microscopy
SLS	Static Light Scattering
SMEDDS	Self-Micro-Emulsifying Drug Delivery Systems
SNP	starch Nanoparticle
SPI	Soy Protein Isolate
TDD	Transdermal Drug Delivery
TDS	Traditional Delivery Systems
TEGDMA	Triethylene Glycol Dimethacrylate
TG	Tragacanth Gum
TGFβ	Transforming Growth Factor Beta
THF	Tetrahydrofuran
TME	Tumor Microenvironment
TNBC	Triple-Negative Breast Cancer
TOC	Tocopherol O-Carboxymethyl
TPP	Tripolyphosphate
TSP	Tamarind Seed Polysaccharide
XG	Xanthan Gum
XRD	X-Ray Diffraction

Foreword

It is with great pleasure that I write the foreword for the book "Graft Copolymers of Polysaccharides for Drug Delivery Systems," edited by Dr. Rishabha Malviya and Miss Sonali Sundram. This comprehensive collection gathers a plethora of information from contributors across the globe for researchers, professionals, and students interested in the connection between polymer science and drug delivery applications.

Graft copolymers of polysaccharides have shown remarkable expansion in recent decades due to the demand for innovative ways to deliver drugs, better therapeutic efficacy, and advanced biomedical technology. This book presents a wide range of information, from the design and synthesis of polysaccharide graft copolymers to their biomedical and pharmaceutical applications.

The book "Graft Copolymers of Polysaccharides for Drug Delivery Systems" is a comprehensive resource that explores the applications of polysaccharide graft copolymers in the preparation of drug delivery systems. The book is a valuable resource for academicians, pharmaceutical experts, and industrial scientists, focusing on the diverse applications of polysaccharide-based graft copolymers to address the challenges. The book looks at the fundamental concepts of creating graft copolymers for specific applications, as well as the potential of grafted polysaccharides in gene transfer, controlled release, and modified drug delivery, widening the frontiers of therapeutic interventions. Methods to develop the graft copolymers, with advantages and limitations described in the book, help researchers select the suitable methods as per requirement. Grafting leads to an overview of tunable materials essential for targeted drug delivery. The editors, Dr. Rishabha Malviya and Miss Sonali Sundram, and the authors have been praised for their commitment to the progression of pharmaceutical sciences.

I'd like to congratulate Dr. Rishabha Malviya, Miss Sonali Sundram, the editors, and the authors for contributing their expertise and enthusiasm to organizing this important source of information. Their commitment to advancing pharmaceutical sciences is clear in the rich content and innovative concepts presented in the chapters.

—Dr. Dhruv Galgotia
CEO
Galgotias University, Greater Noida, Uttar Pradesh, India

About the Book

"Graft Copolymers of Polysaccharides for Drug Delivery Systems" is an in-depth book that focuses on polysaccharide graft copolymers and their significance in reshaping drug delivery systems. The book explores the challenges and potentials of polysaccharide graft copolymers, highlighting the importance of determining the prospects of drug delivery. It presents the foundation for the investigative efforts and cooperative dedication of several experts in the field.

The book discusses the deep concepts of developing polysaccharide graft copolymers for drug delivery applications, as well as the exploration of their use as innovative pharmaceutical excipients. Skilled professionals present their knowledge, delving into up-to-date findings, resolving obstacles, and highlighting advances in polymer derivatives for drug administration. These discussions extend to the cutting-edge delivery systems for gene transfer, highlighting the viewpoint of grafted polysaccharides in drug delivery science.

The book surveys the swiftly evolving field of drug delivery approaches, providing an extensive overview of the potential, bounds, and future opportunities of grafted copolymers. It is recommended for academics, pharmaceutical professionals, and industrial experts to learn more about the transforming function of polysaccharide graft copolymers in drug administration.

Preface

Advancements in polysaccharide-based graft copolymers in drug delivery systems are extensively discussed in the book "Graft Copolymers of Polysaccharides for Drug Delivery Systems." It has been shown that the use of graft copolymers of polysaccharides may effectively address drug delivery issues and provide customized solutions for a range of therapeutic requirements. In addition to discussing the role of polymer derivatives in improving drug delivery efficacy, the book also covers the fundamental principles of designing polysaccharide graft copolymers for drug delivery applications, as well as the most recent developments and difficulties in applying grafted copolymers in innovative drug delivery systems. It also discusses the use of grafted polysaccharides in gene delivery, advancements in modified drug delivery with Prosochit®, and developments in controlled drug delivery with grafted copolymers. The chapter offers a combination of theoretical frameworks, experimental methodology, and practical consequences, giving readers a complete understanding of the growing phase of polysaccharide graft copolymers in drug delivery systems. This book is meant to provide a roadmap for researchers and industry professionals exploring the complex field of polysaccharide-based graft copolymers, enabling innovations and essential advancements in the field of drug delivery systems.

CHAPTER 1

Designing Polysaccharide Graft Copolymers for Drug Delivery Applications

ARISTEIDIS PAPAGIANNOPOULOS,[1] THEODORE SENTOUKAS,[2] EFSTATHIA TRIANTAFYLLOPOULOU,[3] STERGIOS PISPAS,[1] and NATASSA PIPPA[3]

[1] *Theoretical and Physical Chemistry Institute, National Hellenic Research Foundation, Athens, Greece*

[2] *Center of Polymer and Carbon Materials, Polish Academy of Sciences, Zabrze, Poland*

[3] *Department of Pharmaceutical Technology, Faculty of Pharmacy, Panepistimioupolis Zografou, National and Kapodistrian University of Athens, Athens, Greece*

ABSTRACT

One of the key advantages of polysaccharide graft copolymers is their biocompatibility. Graft polymerization enhances polysaccharides by preserving their beneficial properties while introducing new functionalities. Advances in controlled radical polymerization and ring-opening polymerization (ROP) techniques have opened new synthetic pathways for creating polysaccharide graft copolymers with precise architectures, utilizing a wide range of functional monomers. The micellar self-assembly and complexation of grafted polysaccharides, as well as the colloidal properties of their aggregates, will be examined using physicochemical experimental methods. Morphology and stability in aqueous media are crucial for the application

Graft Copolymers of Polysaccharides for Drug Delivery Systems.
Rishabha Malviya and Sonali Sundram (Eds.)

of these synthesized nanoparticles, whether they are unloaded or loaded with drugs, proteins, or polynucleotides. This chapter will also explore the use of polysaccharide graft copolymers in pharmaceuticals, where various types of nanoparticles have been employed for the delivery and controlled release of active pharmaceutical substances. Additionally, targeting specific tissues can be achieved using these nanoparticulate systems. Polysaccharide graft copolymers have also garnered attention as protein and non-viral gene delivery vectors due to their favorable properties. The aim of this chapter is to summarize the primary methods used for synthesizing and characterizing polysaccharide graft copolymers, while highlighting their pharmaceutical applications, advantages, and challenges as nanocarriers.

1.1 INTRODUCTION

Polysaccharides are natural, biodegradable, and biocompatible biomaterials that have been widely used as drug delivery carriers due to their unique properties. However, their low water solubility, poor mechanical strength, and limited functional groups for chemical modification have restricted their practical applications. Nevertheless, polysaccharide graft copolymers have gained popularity in biomedicine due to their intrinsic biochemical and biophysical qualities, such as biocompatibility, degradability, extended plasma circulation time, and targeted delivery. Recent advances in polysaccharide drug vectors and a deeper understanding of the natural functions of polysaccharide graft copolymers suggest that these carriers may offer advantages over synthetic material-based alternatives in terms of efficacy, safety, colloidal stability, circulation time, and ease of manufacturing [1–4]. Graft copolymerization involves the attachment of one or more types of monomers to the backbone of the polymer chain. In other words, it is a technique to modify polysaccharides by introducing hydrophilic, hydrophobic, or functional groups, or polymeric chains onto the main polysaccharide backbone. This approach enables the synthesis of hybrid natural-synthetic copolymers with enhanced physicochemical properties, such as improved solubility, stability, and drug loading capacity, which are critical for drug delivery applications [1–4].

Many techniques, such as free radical polymerization or others, can start this process [5, 6]. In the process of creating graft copolymers using free radical graft copolymerization, for instance, the polymer backbone is first functionalized with free radical centers. Depending on the type of monomer and the grafting density, the resulting copolymers display a variety of hybrid features, such as enhanced mechanical strength, thermal stability, responsivity to pH, temperature, and ionic strength, as well as amphiphilicity [5, 6]. Graft copolymers are used in many different industries, such as electronics, coatings, and biomedicine. Natural polysaccharides, that is, alginate, dextran, and others, are some of the most widely studied polysaccharides for drug delivery and targeting applications. These polysaccharides have been modified via graft copolymerization with various monomers, such as acrylic acid (AAc), methacrylic acid (MA), N-vinylpyrrolidone (NVP), polyethylene glycol, and poly (lactic-co-glycolic acid) (PLGA), to form copolymers with improved properties. These copolymers have been used to prepare different types of drug delivery systems (DDSs), including nanoparticles, microparticles, hydrogels, and films [1–4]. The polysaccharide graft-copolymers are biocompatible materials, and there are numerous manuscripts in the literature for drug/protein delivery and tissue engineering because they have shown excellent results *in vitro* and *in vivo*. Some authors declare that they may serve as a new generation of biomaterials for pharmaceutical and biomedical applications [7].

The scope of chapter one is to summarize the main methods that are utilized to synthesize and characterize polysaccharides graft copolymers, as well as to highlight their pharmaceutical applications, advantages, and challenges as nanocarriers. Different nanoparticles that are self-assembled from polysaccharide graft-copolymers will be presented in terms of their physicochemical, morphological, and biological characteristics, as well as for their ability to encapsulate Active Pharmaceutical Ingredients (APIs), that is, low molecular weight (LMW) APIs to proteins/peptides and nucleic acids.

1.2 SYNTHESIS OF POLYSACCHARIDE GRAFT COPOLYMERS

Polymerization methods have evolved through the years to become important tools for the synthesis of complex architectures from a great variety of monomers in different solvents. Scientists have also managed to find new "tricks" in order to prepare bioconjugate-nanostructures that are comprised of (co)polymers and biological molecules, such as proteins or DNA/RNA, by simply utilizing them as starting points/initiators/mediators for polymerization of the selected monomers. In this manner, many groups have utilized grafting techniques to prepare (co)polymer-polysaccharides nanostructures, in order to achieve the best of both worlds [8, 9]. Grafting techniques can be mainly categorized as "grafting from," "grafting onto," and "grafting through." "Grafting through" occurs when the already polymerized sidechains work as starting points for the polymerization of the backbone. This method allows the synthesis and characterization of well-defined polymeric chains that will be grafted on the forming backbone. On the other hand, it suffers from random copolymerization between backbone monomers and polymerized chains and low conversion rates. "Grafting onto" is the coupling of polymerized sidechains onto the preformed backbone via its reactive side-groups. Well-defined grafts can be prepared by this method since both sidechains and backbone can be meticulously synthesized and characterized. Nevertheless, the coupling technique must be carefully chosen, and several purification steps are needed for separating unreacted species. "Grafting from" refers to the use of the synthesized backbones' functional pendant groups as starting points for the polymerization of the selected monomer(s) as side chains. This technique has been utilized for living radical polymerizations after modifying the backbone in such a way as to play the role of the macroinitiator/polymerization agent and provide control over the final macromolecular architectures [9–13].

Chemical modification is needed for the polysaccharides to become macroinitiators for further grafting polymerization schemes. Polysaccharides bear functional pendant groups such as carboxyl (hyaluronic acid, alginic acids), amine (chitosan), or hydroxyl groups (starch, cellulose) that can be further modified in order to be converted into starting points for polymer grafting. In this context, common chemical reactions have been proposed by researchers, such as esterification, etherification, oxidation, amination, and amidation, to name a few, depending on the grafting and polymerization method that will be used [9, 13, 14]. Radical polymerization techniques have been established for the synthesis of polysaccharide graft copolymers since it is an easy, versatile, and quick method to produce aimed macromolecules. Traditional free-radical to more controlled techniques, such as RAFT, ATRP, and ROP, have been utilized depending on the aim of the synthesis. The common characteristic of all these techniques is the formation of an active site (e.g., radical) from a chemical compound and its propagation through monomers, that are forming the polymer chains through addition, one unit after another [8, 9].

1.2.1 RING OPENING POLYMERIZATION (ROP)

ROP is another polymerization technique that makes use of anionic, cationic, or radical reactive centers that act as a polymerization initiator for ring-shaped monomers [15]. It has been excessively used for the grafting of ε-caprolactone, oxazolines, PEG, PTHF, and more into polysaccharides and biomolecules in general. The idea is rather simple. The polysaccharide is modified in a chemical way, which produces an active group that will attack the cyclic monomer and propagate the polymerization. Youssouf et al. [16] have grafted polycaprolactone from carrageenan oligosaccharides via a four-step procedure. Carrageenan was initially depolymerized using a specific enzyme into oligo carrageenans of 8 units after a 6-hour hydrolysis process. Then, the hydroxyl groups of the oligo carrageenans were partially acetylated at a 60% degree, while the remaining free ones were utilized for the polymerization of PCL via ROP, with the use of Sn $(Oct)_2$ as a catalyst. Different PCL lengths were obtained by modifying the monomer-to-initiator ratio. Restoring the hydroxyl groups via deacetylation leads to amphiphilic caprolactone-oligocarrageenan graft copolymers. Zuppolini et al. [17] have utilized cellulose type 20 to prepare graft copolymers with PCL side chains via ROP. Sn $(Oct)_2$ was used as a catalyst inside an ionic liquid ([Bmim] Cl) medium. In this way, cellulose was completely soluble, and its grafting points were accessible to ε-caprolactone for successful polymerization. The group prepared Cell-g-PCL in a ratio of 1:15 to ensure the existence of amphiphilic character that will lead to interesting self-assembly abilities for further applications.

Abbasi et al. [18] hydrolyzed medium-M_w chitosan (85% deacetylated) with 85% phosphoric acid at 70°C in order to receive lower M_w. As a second step, m-PEG propionaldehyde was synthesized by oxidation of PEG's hydroxyl group in order to produce stable linkers for grafting PEG onto chitosan. PEGylated medium and low M_w chitosans were prepared via $NaCNBH_3$ into aqueous acetic acid and methanol. As a final step, palmitic acid was introduced to the system as a hydrophobic component in order to produce amphiphilic self-assembled nanoparticles. Carvalho et al. [19] synthesized an alkyne-terminated PCL and an azide-functionalized pullulan with a molecular weight of 120 kDa. In the second step, the group synthesized pullulan-PCL graft copolymers via the "grafting onto" technique utilizing click chemistry coupling. Nishimura et al. [20] have synthesized an azide-functionalized PPO of 2 kDa via anionic polymerization in the presence of 6-azide-1-hexanol and phosphazene base P_4-tBu. Pullulan of 130 kDa was alkyne-functionalized with 5-hexynoic acid in the presence of DCC and DMAP. Later on, the group synthesized pullulan-g-PPO via the "grafting onto" technique by a copper-catalyzed Huisgen reaction between azide-functionalized PPO and alkyne-modified pullulan. Deng et al. [21]

have utilized the "grafting onto" method by initially preparing the poly (THF-co-ECH) chains via living cationic ROP. The newly synthesized living chains were added along HPC via a new reaction that resulted in grafting onto the OH groups of HPC with controlled termination. As a final step, the group conducted a quaternization of the chlorine groups of ECH to obtain quaternized/cationic branches.

Gao et al. [22] have prepared polysaccharide graft copolymers with the "grafting onto" technique using PGA and SA. Acylation of both PGA and SA was performed as a first step, followed by the synthesis of the PTHF chains via AllylBr initiator and $AgClO_4$. The final step refers to the condensation reaction and nucleophilic substitution of the PGA's hydroxyl groups with the PTHF chains. Zhao et al. [23] have grafted poly (IB-THF) onto dextran in a multiple-step procedure. The first step refers to the partial alkylation of dextran in order to protect some of the OH groups. As a second step, the synthesis of PIB-allylBr was achieved, followed by copolymerization of THF. Then, the P(IB-THF) chains were grafted onto the unprotected hydroxyl groups of dextran. Lavikainen et al. [24] have synthesized gellan gum-poly(2-ethyl-2-oxazoline) grafts via the "grafting onto" method. The first step was the polymerization of EtOx via cationic ROP. Chemical modification of GG led to the removal of the proton from hydroxyl groups. As a final step, the coupling reaction between the pEtOx and GG leads to the grafting of the polymer onto the O^- groups of GG.

1.2.2 REVERSIBLE ADDITION-FRAGMENTATION CHAIN TRANSFER POLYMERIZATION (RAFT)

RAFT polymerization is another method that has been utilized to synthesize polysaccharide graft copolymers. It is a pseudo-living/controlled polymerization method that utilizes an organic, small chain transfer agent molecule (CTA) to mediate the polymerization. It has one characteristic Z and one characteristic R group that must be wisely chosen in order to keep the radicals in equilibrium and maintain the control of the polymerization. RAFT can be applied to all monomers that can be polymerized via free-radical polymerization in a wide selection of solvents and temperatures [25]. RAFT polymerization can be utilized to create well-defined grafts from polysaccharides. The notion is to modify the polysaccharide into a macroCTA via common chemical reactions. Ikkene et al. [26] have converted dextran into a macro-CTA group through many steps of oxidation, amidation, and amination, utilizing 4-(propylthiocarbonothioylthio)-4-cyanopentanoic acid. The resulting macro-CTA had 12 starting points for further polymerization. Thus, HPMA was then grafted onto dextran-CTA via RAFT PISA photopolymerization in aqueous media, using TPO-L as the photoinitiator [27].

Hatton et al. [28] introduced amine groups to the reducing chain end of xyloglucan of 17.5 kDa through reductive amination. The resulting XG-NH_2 was then reacted with 4-cyano-4-(phenylcarbonothioylthio) pentanoic acid to prepare a macro-CTA. The coupling reaction took place between the NH_2 terminated group with the carboxylic group of the CTA. MMA was further polymerized via RAFT polymerization to obtain XG-g-PMMA with three different degrees of polymerization for PMAA. Ferji et al. [29] are the first to have prepared polysaccharide graft copolymers via PISA RAFT in aqueous media. The group has utilized dextran as the backbone for grafting PHPMA chains via RAFT photopolymerization. The group initially synthesized the TTC CTA, which was then coupled by partial esterification with the OH groups of dextran using CDI. The dextran-TTC macroCTA was then utilized for the photopolymerization of HPMA, providing good control and well-defined polymeric side chains. Pilipenko et al. [30] have utilized heparin and chondroitin sulfate as backbones to prepare Hep-g-PNIPAAm and CS-g-PNIPAAm nanoparticles. The group modified the hydroxyl pendant groups of both Hep and CS with Boc-EtNH_2 in order to prepare the coupling with DDMAT via esterification and synthesize a Hep-CTA and CS-CTA for grafting NIPAAm from their CTA-side groups.

1.2.3 ATOM TRANSFER RADICAL POLYMERIZATION

ATRP is a living and controlled reversible-deactivation radical polymerization technique that makes use of an alkyl halide as the initiator and a transition metal/ligand as a catalyst. The dormant species is activated by the catalyst to create free radicals. The catalyst becomes oxidized, its oxidation state is increased, and the reaction equilibrium is shifted towards the formation of low-concentration radicals, reducing the chances of radical-radical terminations. The number of chains is directed by the number of initiator molecules, which have the same probability of polymerization, hence providing good control and well-defined polymer chains.

ATRP can work with all monomers that can be polymerized with free radical polymerization in a wide variety of solvents and temperature ranges [31]. Polysaccharides graft copolymers can be rather easily synthesized using ATRP, and the idea is to convert the polysaccharide of interest into a macro-initiator for ATRP in order to mediate the polymerization of other monomers. Yuan et al. [32] have modified ethyl cellulose with 2-bromoisobutyryl bromide, resulting in a replacement of hydroxyl groups by bromide. The EC-Br was further utilized as the initiator for the synthesis of two different graft copolymers comprising EC-g-PDMAEMA and EC-g-P(MEO_2MA-co-DMAEMA), with CuBr/PMDETA as the ATRP catalyst. Ferji et al. [23] have converted a dextran polysaccharide into an ATRP macro-initiator within a four-step procedure. Initially, the OH groups were partially acetylated, while the remaining ones were esterified with 2-bromoisobutyryl bromide. PDEGCholA was polymerized via ATRP as hydrophobic grafted blocks. After the polymerization, the acetylated groups were deprotected back to hydroxyl groups via deacetylation.

1.2.4 FREE RADICAL POLYMERIZATION

Free radical polymerization is one of the simplest methods for polymer synthesis but suffers from random termination of radicals and uncontrolled chain growth that results in polymers with broad molecular mass polydispersity. It is based on the production of free radicals by a chemical compound (typically an azo compound), most frequently caused by heating or radiation, that polymerizes the active double bond site of a vinyl monomer [34]. Free radical polymerization has been used for the synthesis of polysaccharide graft copolymers since it is an easy and quick method. Wen and Oh [35] have prepared two separate polysaccharide graft copolymers, based on carboxymethyl cellulose sodium salts (CMC) via aqueous free-radical polymerization technique. P(MEO_2MA-*co*-OEGMA) was initially grafted on a CMC of 250 kDa. The same monomers were used again with the addition of DMA to produce crosslinked P(MEO_2MA-*co*-OEGMA-*co*-DMA). Ortiz et al. [36] have prepared a macro-initiator backbone of maleoylagarose through the esterification of agarose and maleic anhydride. The group further grafted NIPAAm onto the maleoylagarose via a simple free radical polymerization. Liu et al. [37] have synthesized alginate-PNIPAAm grafts via the "grafting onto" technique. First, NIPAAm polymerization was mediated, utilizing CAH as a CTA. The amino-PNIPAAm was then purified and coupled with alginate using EDC and NHS.

1.3 EVALUATION OF NANOPARTICULATE SYSTEMS BASED ON POLYSACCHARIDE GRAFT COPOLYMERS

Self- and co-assembly of macromolecular compounds and especially block copolymers in aqueous media are very important processes that define the formation of multifunctional nanostructures [38, 39]. Polysaccharide graft copolymers may self-assemble in various ways depending on the physicochemical properties of their components (Figure 1.1). Dynamic light scattering (DLS) is very sensitive in the detection of the formation of aggregates and in any alteration of the hydrodynamic size distribution of species in solution. Static light scattering (SLS) is used to measure the mass and gyration radius of scattering particles, and electrophoretic light scattering (ELS) is used to quantify their surface potential. Microscopy techniques may complement the light scattering methods by providing information on the shape and morphology of the nanostructures. Small-angle scattering (SAS) techniques, like SANS and SAXS, are advanced methods for the shape and internal structure characterization (Figure 1.1) of self-assembled aggregates. Therefore, the aforementioned methods have been proven very important in the studies of polysaccharide graft copolymers in solution.

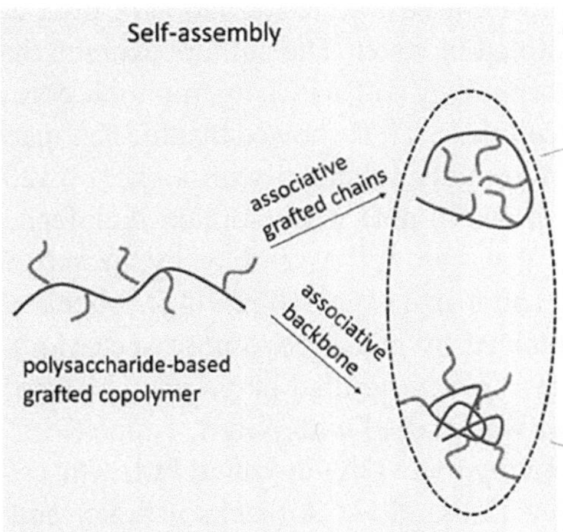

FIGURE 1.1 Simplified illustration of self-assembly in polysaccharide graft copolymers and experimental methods for the investigation of their properties.

Block terpolymer chains of PDMAEMA–PCL–mPEG (poly(dimethyl aminoethyl methacrylate)-polycaprolactone-poly(ethylene glycol)) were grafted on tragacanth gum (TG) to create a doubly stimuli-responsive polysaccharide-based graft copolymer. The PDMAEMA component introduced the pH and temperature responsiveness to the copolymer. PCL is a hydrophobic polymer, and PEG is a hydrophilic one. The system was used for the sustained release of the hydrophobic drug quercetin. These copolymers formed micelles in aqueous solutions with sizes between 90 and 100 nm, which increased upon loading of the model drug. The size in solution appeared to increase abruptly at 45°C as the PDMAEMA block became strongly hydrophobic and the micelles assembled into larger aggregates. In addition, the decrease in pH led to a decrease in the micellar size because of the weak cationic charge of the PDMAEMA [40]. The hydrophilic polysaccharide maca (*Lepidium meyenii* Walp.) was grafted on a hydrophobic backbone of PLGA. This copolymer tends to form core-shell micelles with a hydrophobic core of self-assembling PLGA segments and a hydrated shell of the polysaccharide chains [41]. DLS was used to evaluate the size distribution in the PBS buffer. Narrow size distributions were found with an average hydrodynamic diameter of about 120 nm. The size increased upon the addition of glutathione (GSH) because of the redox-sensitive disulfide grafting bond. TEM revealed the spherical morphology of the micelles and the formation of random aggregates upon the addition of GSH.

Amphiphilic copolymers with grafted chains of poly(ε-caprolactone) (PCL) on a cellulose backbone were used for the formation of drug carrier nanoparticles as they self-assembled in water. The authors claimed that the nanostructures had a micellar morphology with a PCL hydrophobic core and a shell consisting of amphiphilic cellulose. TEM showed that the nanoparticles formed by electrospraying were roughly spherical with a size of 320 nm. The nanoparticles became larger (~390 nm) when sodium diclofenac was encapsulated. The nanoparticles that were self-assembled by the antisolvent technique were aggregated and had a size of 30–60 nm [17]. Micelles with a polyelectrolyte shell were obtained by mixing two polysaccharides with different grafted macromolecules. In both grafted polymers, chitosan (CS) was the backbone. One polymer was grafted with poly-L-lysine (PLL) and the other with anionic citraconyl anhydride (CA)-modified PLL. The cationic PLL blocks interacted strongly with the PLL-CA blocks in water and self-assembled in a neutralized core, whereas the CS chains formed the charged micellar corona. At pH 7.4, the micelles had well-defined size distributions with mean diameters of about 55 and 76 nm, as measured by TEM and DLS, respectively. Their zeta potential was higher than 20 mV as it was measured by ELS. The difference in the observed sizes was attributed to the increased hydration of the micelles in solution (DLS experiments) in contrast to the conditions of the TEM experiment. The micelles were pH-responsive as they were disassembled at pH 5, and their size dropped to about 35 and 23 nm, respectively, as measured by TEM and DLS. The zeta potential increased to almost 30 mV. The reason for these changes was the hydrolysis of CA, which turned the PLL-CA chains cationic. The hydrolysis of CA at pH 5 retrieved the PLL amines that were protected in the core at pH 7.4 and produced the antimicrobial activity of the graft copolymers on *E. coli* [42].

The widely investigated cationic polysaccharide chitosan (Chi) has been grafted with quaternized cyclodextrin (QCD) in the graft copolymer QCD-g-Chi. The positively charged copolymer created complexes in aqueous media via electrostatic interaction with the anionic extracellular polysaccharide hyaluronic acid (HA). The complexes had a size that was of the order of 500 nm and increased as the molar mass of HA increased. The surface charge of the complexes was positive and decreased by the increase of added HA, while this decrease was stronger for the high molar mass HA. HA was used for its properties to hydrate the skin and normalize keratinization, while its inclusion in the complexes prolonged its release *in vitro* in comparison to the free HA. TEM showed that the QCD-g-Chi polymers created spherical core-shell micelles in solution where the core consisted of the hydrophobic QCD heads and the shell contained the hydrophilic CS backbone. HA adsorbed in the outer periphery of the micelles and was in a contracted conformation [43]. Chi grafted with poly (ethylene glycol) (PEG) in combination with poly(styrene sulfonate) (PSS) was used for the preparation of layer-by-layer nanocapsules on magnetite nanoparticle cores for the immobilization of BSA in the outer periphery. The morphology and size of the nanocapsules were characterized by TEM, AFM, and DLS. The stability of the prepared nanocolloids was attributed to their hydrophilic coating [44].

In another work, CS was grafted with a copolymer of methacrylamide (MA) and acrylonitrile (AN) (Chi-g-poly (MA-co-AN) HA) and was electrostatically complexed with HA. X-ray diffraction (XRD) showed that grafting Chi to poly (MA-co-AN) and complexation with HA removed Chi's ability to crystallize. These complexes were used to fight against the biofilm formation of *Candida albicans*. Scanning electron microscopy (SEM) revealed that the complexes had a smooth and homogeneous surface morphology [45]. A system based on CS oligomers grafted with the negatively charged 2,3-dimethyl maleic anhydride (DA) could be used as a paradigm for application in polymer analogs. The negatively charged DA was used to electrostatically complex with the positively charged antibiotic polymyxin B. In an acidic environment, the amide bonds between Chi and DA were hydrolyzed, leading to the release of the drug [46]. HA grafted with poly(ethylene imine) (PEI) was synthesized to create core-shell nanovehicles combined with Au nanoparticles that would be able to deliver antitumor drugs. A core of electrostatically bound PEI and DNA was the basis of the formulation. The size of the nanoparticles was about 300 nm, and it decreased with the increase of the Au content. At the same time, the negative zeta potential increased to zero. This was attributed to the HA chains that remained on the outer periphery of the nanoparticles [47]. HA has also been grafted by thermoresponsive poly(N-isopropyl acrylamide) PNIPAM to prepare self-assembling nano gel nanocarriers of curcumin. The grafted macromolecules self-assembled in aggregates in aqueous media at room temperature. Above the LCST, the curcumin was released from the nanogels. The size of the loaded nanogels was 100–300 nm, and their shape was spherical, as found by DLS and TEM [48].

An amphoteric grafted copolymer with a cashew gum (CG) backbone and poly(O-methacryloyl-l-serine) (poly (SerMA) sidechains was synthesized as a flocculant of kaolin particles in aqueous media. The effect of grafting on the conformation of CG, that is, the increase in the molecular size, was quantified by intrinsic viscosity measurements [49]. Kaolin particles are positively charged at low pH and have an isoelectric point pI of 3. The interaction between the grafted polymer and the clay particles was found at pH 3.5, as revealed by turbidity measurements. The improvement of the turbidity removal at this pH range was attributed to the protonation of the NH_2 groups of SerMA and the interaction with the negatively charged kaolin. Rice husk-based based cellulose (RH) was grafted with β-CD by either epichlorohydrin or glutaraldehyde [50]. The grafting was confirmed by FTIR, where the signatures of the two different macromolecules were observed. In addition, TGA experiments detected the decomposition of both cellulose and β-CD and the dehydration of free water from ungrafted and grafted RH. The two grafted copolymers had good efficiency in the remediation of heavy metal contaminated water, as it was indicated by adsorption performance on Pb (II) cations.

A grafted copolymer with a carboxymethylcellulose (CMC) backbone showed thermoresponsive self-assembly due to its poly(ethylene oxide/propylene oxide) sidechains [51]. At room temperature, the copolymers exhibited no significant self-assembly, and the radius of gyration and molar mass of pure CMC and grafted CMC were obtained by SLS. The presence of a cloud point was investigated by UV-Vis absorption experiments at different temperatures. The graft copolymers showed no cloud point temperature (T_{CP}) in water in contrast to the linear polyether amines. This was explained by the polyelectrolyte nature of the backbone, whose stretched conformation did not allow the side chains to interact strongly. However, upon the addition of salts, a T_{CP}, which decreased as a function of salinity, was observed. This was attributed to the screening of electrostatic interactions between the segments of the CMC backbone and the salting out effect that reduced the polymer-solvent attraction for the polyether amines. DLS experiments at 25 and 60°C in water, 0.5 M NaCl, 0.5 M K_2CO_3, and synthetic seawater quantified these effects further. In aqueous solutions with no added salt, the temperature increase caused a small decrease in the aggregates of the copolymers from 130 to 110 nm for $PEOPPO_{2070}$ (low grafting density/high molar mass graft) and from 160 to 140 nm for $PEOPPO_{600}$ (high grafting density/low molar mass graft). The sizes were much higher in the absence of salts than in the presence of salts, which was attributed to the shrinkage of the backbone of the graft and the intermolecular hydrogen bonding. The increase in temperature was accompanied by the shrinkage of individual polymer chains and formation of large aggregates because of the appearance of intra- and inter-chain hydrophobic interactions, respectively.

SAS methods, namely small angle X-ray scattering (SAXS) and small angle neutron scattering (SANS), can be effectively used to reveal the morphology of self-assembling polymeric objects at the nanoscale. These methods provide spatial resolution in the range from 1 to 100 nm in a non-invasive manner and by averaging over a vast number of scattering particles (macroscopic sample volume). Amphiphilic graft copolymers of dextran with hydrophobic grafted side chains of poly (2-hydroxypropyl methacrylate) (DEX-g-PHPMA) were studied by SAXS and were illustrated to form vesicular nanostructures in aqueous solutions. SAXS data analysis proved the spherical shape of the vesicles consisting of a membrane with a DEX hydrophilic exterior and an HPMA hydrophobic interior. SAXS also quantified their radius (10–20 nm) and membrane thickness (20–30 nm) that decreased and increased, respectively, as a function of the PHPMA degree of polymerization (DP = 100–200) at a constant number of grafted chains. At the highest degree of polymerization of PHPMA (DP = 300), the self-assembly led to spherical nanoparticles with a PHPMA hydrophobic core and a DEX hydrophilic shell. These results were complemented by DLS, cryoEM, and atomic force microscopy (AFM) experiments [26]. The case of grafted PHPMA with DP=400 was investigated in a separate study. The authors showed by DLS and cryo-TEM that different morphologies (spherical micelles and worm-like micelles) could be obtained by different preparation methods, that is, photo-polymerization-induced self-assembly, nanoprecipitation, conventional self-assembly, and temperature conditions

[52]. SAS has also been applied to investigate the self-assembly of another graft polysaccharide. In more detail, pullulan (Pul) grafted with the thermo-responsive polymer PPO was investigated. SAXS profiles were collected from 4°C to 25°C at small intervals of 1–2°C. At low temperatures, the form factor of a thin cylinder represented the data with a disc radius of 19 nm, a 3 nm thickness of a hydrophobic core, and a hydrophilic shell of 0.5 nm. At higher temperatures, the cylinders transformed into spherical vesicles of a diameter of 45 to 70 nm, which was an increasing function of polymer solution concentration [20].

In another system, hydrogels were formed by the association of modified gellan with surfactant worm-like micelles. The gellan gum chains were grafted with cholesterol groups while the grafted copolymers were immersed in solutions that contained a mixture of the co-surfactants polyoxyethylene cholesteryl ether ($ChEO_{10}$) and triethylene glycol monododecyl ether ($C_{12}EO_3$). The two components (grafted polymer and surfactant worm-like micelles) resulted in complex fluids of predominantly solid-like behavior. The surfactant micelles interacted with the modified gellan to form a percolating network above a threshold concentration. The $ChEO_{10}$ micelles had a disc shape with a diameter of 14 nm and a thickness of 4 nm, as revealed by SANS. The addition of the co-surfactant led to their elongation and the formation of worm-like micelles. SANS also revealed that while the shape and size of the surfactant micelles were unperturbed by the presence of the polymer, their spatial arrangement was affected. Micellar clusters were formed as the polymer induced inter-micellar bridging [53].

1.4 DELIVERY OF APIS

Dual temperature/acidic pH-responsive nanogels composed of CMC were used for the delivery of doxorubicin. Grafted oligo (ethylene oxide)--based copolymers with thermoresponsive properties were designed to modulate the release of the loaded anticancer API. The doxorubicin release was faster in temperatures higher than the LCST and in acidic pH. The last observation is very important due to the acidic environment inside the tumor cells. The prepared system also exhibited stability in the presence of blood proteins. Additionally, *in vitro* cytotoxicity experiments showed that the bio-nano gel loaded with doxorubicin is a suitable system for anticancer therapy due to the significant decrease in cell viability of HeLa cells [35].

The alginate-g-polyvinylpyrrolidone copolymer was used for the encapsulation of ibuprofen, which is a nonsteroidal anti-inflammatory molecule with several applications for the treatment of musculoskeletal problems. Alginate-g-polyvinylpyrrolidone copolymer solution containing ibuprofen was cross-linked by glutaraldehyde for the preparation of beads with pH-responsive properties. The API/copolymer ratio ranged from 1/8 to 1/2, while the encapsulation efficiency was higher than 90% in all formulations studied. As the N-vinyl-2-pyrrolidone graft yield, the ibuprofen release profile was also increased up to 8h in a pH-dependent manner (with a zero release at the pH of the stomach and with a non-Fickian transport at the pH of the small intestine) [54]. Buflomedil hydrochloride was encapsulated into tablets composed of acrylamide grafted copolymer of locust bean gum. The oral acute toxicity studies showed that this system was biocompatible. Biodegradability was also investigated, including complete degradation of the formulation at 21 days. A prolonged release of the encapsulated API was obtained (until 10 hours) in comparison to Hydroxy Methyl Cellulose (HPMC) tables. All the API was released until 24 h [55]. In the same context, agar was modified by microwave-assisted grafting with acrylamide monomer, resulting in poly acrylamide grafted agar for the controlled release of 5-amino salicylic acid. The authors investigated how the percentage grafting affected the t_{50}. They observed that sustained release was achieved with a higher percentage of grafting. The release of the API from this grafted agar matrix was near zero in acidic pH and increased as pH turned into neutral values [56].

Poly(ε-caprolactone)-g-dextran was used for the encapsulation of doxorubicin. The encapsulation was achieved by the nanoprecipitation method, and the resulting micelles exhibited size at 100 nm with low polydispersity. The biodegradable "reverse" oligosaccharide-based amphiphilic graft copolymer morphology with a hydrophobic backbone and hydrophilic side chains was found to be ideal for sustained release of this anticancer API in a pH-dependent manner. The effectiveness of poly(ε-caprolactone)-g-dextran reversed micelles was evaluated in two different cell lines (colorectal cancer cells and healthy human fibroblasts), showing a significant degree of internalization in cancer and low toxicity in healthy cells [57]. Dextran-graft-poly (N-isopropyl acrylamide) copolymers were formed via Schiff base synthesis for the encapsulation of doxorubicin. The nano vectors were prepared by pH-gradient protocol with particle size <200 nm, ideal for passive targeting. The release of the anticancer API was higher in acidic pH with kinetics close to the Ritger-Peppas model. The release of doxorubicin at neutral pH fitted the mechanism of Fickian diffusion. The doxorubicin-loaded nanosystems exhibited a higher cytotoxicity profile in cancer cells in comparison to healthy cell lines, an observation which was in line with the invitro release assay that showed a higher % of API released at pH = 5.5. According to the cellular uptake assay, the prepared dextran-graft-poly(N-isopropyl acrylamide) nanoparticles were captured via the mechanism of endocytosis, and their distribution in the cytosol was higher, while the free API diffused across the plasma and nuclear membrane and released in a sustained/controlled profile responding to the acidic pH of endosomes [58].

The amphiphilic copolymer of CG grafted poly(ε-caprolactone) was used for the preparation of nanoparticles for the loading of the anticancer agent epirubicin. The nanoparticles loaded with the API exhibited a size of around 300 nm (increased size in comparison to the pure ones), with great colloidal stability probably to negative zeta potential values. Murine fibroblast non-tumor cells, human breast adenocarcinoma cells, and human colorectal cancer cells were used for the cytotoxicity assay, showing increased uptake of the API in the cell lines. The morphology of the cell lines changed only at the highest concentration of the nanosystems. The colloidal nanosystems showed excellent hemocompatibility (hemolysis percentage below 5%) of the amphiphilic copolymer of CG grafted poly(ε-caprolactone) and additional evidence that the API was encapsulated effectively into the nanoparticles [59]. Another example of polysaccharide graft copolymers is Moringa gum-g-poly(N-vinyl-2-pyrrolidone). Firstly, Moringa gum is a biopolymer composed of galactose, rhamnose, and trace amounts of uronic acid with pendant -OH groups. The synthesized graft copolymer increased the bio-adhesive properties of natural biopolymers, especially gums. The outcomes from the ex vivo bio-adhesion experiments showed that the Moringa gum-g-poly(N-vinyl-2-pyrrolidone) buccal formulations exhibited a higher degree of bio-adhesion due to the synergistic effect of the natural gum and the poly-vinylpyrrolidone. The same results were also observed from the swelling studies. The formulation of clotrimazole-loaded buccal discs/ buccal patches, including other excipients for the amelioration of plasticity and lubricant properties, exhibited sustained release of the API until 24 hours [60]. The synthesis of acylated sodium alginate-g-(polytetrahydrofuran-b-polyisobutylene) terpolymer has also appeared in the recent literature. The loading of the micelles with ibuprofen as model APIs was investigated with prolonged release achieved in two days in a pH-dependent manner in the pH=7.4. The preparation of acylated sodium alginate-g-(polytetrahydrofuran-b-polyisobutylene) terpolymer/Ag composites exhibited antibacterial behavior against Escherichia coli and Staphylococcus aureus. The same biomaterial showed effective protein adsorption, too. The aforementioned properties made them a multifunctional material with several pharmaceutical properties [61].

Additionally, acylated dextran-g-polytetrahydrofuran-graft-copolymer with silver nanoparticles also exhibited bio- and hemo-compatibility and great resistance to protein adsorption, as well as delayed-release properties, too [62]. A cellulose-graft-poly(ε-caprolactone) was synthesized, and the prepared nanomicelles by electrospraying with a size around 50 nm exhibited high encapsulation efficiency and sustained release of sodium diclofenac. Namely, the release kinetics included a burst release of the incorporated API in the first 24 hours and a delayed release in the next six days. The authors declared that the interactions between amphiphilic polymer chains and water molecules governed both the nanoparticle formation and the release of the API [17]. The amphiphilic graft copolymer oxidized carboxymethylcellulose (CMC) with hydrophilic poly(ethylene glycol) (PEG) and dodecylamine (DDA) were prepared with biocompatible and biodegradable properties. The encapsulation of the hydrophobic API nifedipine was achieved by the oil/water emulsion method to improve its solubility and bioavailability. Different micellar formulations were prepared with different compositions of the copolymers. All the micelles exhibited sizes between 100–250 nm with negative zeta potential values and encapsulation efficiencies between around 20%. All the formulations were characterized by good colloidal stability and pH-dependent release profile at neutral pH, mimicking the conditions of the small intestine [63].

Polyurethane grafts on linear dextrin were used to design and develop injectable hydrogels for the encapsulation of the model API dexamethasone (DEX) for the treatment of melanoma. Cellular experiments demonstrated the biocompatibility of brush copolymers. According to cellular uptake studies, the endocytosis of API-tagged contrast agents suggests increased transit of biologically active material into living cells. Furthermore, *in vivo* experiments on melanoma mice showed the effectiveness of the controlled DEX release from the injectable hydrogel with considerable tumor suppression without any adverse effects. Figure 1.2 shows the effect of *in vivo* controlled drug release and its underlying effects on organs and melanoma tumors [64].

The binary grafts Chi-graft-poly (AA-cop-AAm) and Chi-graft-poly (AA-cop-AN) (where Chi is chitosan, AAm is acrylamide, and AN is acrylonitrile) were synthesized with various by varying the composition of AA. The biocompatible and biodegradable properties of these polysaccharide graft copolymers made them suitable excipients as drug delivery carriers. Several mechanical, morphological, and thermotropic studies were carried out showing increased porosity, strength, and thermal stability of the chitosan graft-copolymers. The diclofenac loading was also increased four times in chitosan graft copolymers hydrogels, followed by a sustained release at neutral pH [65].

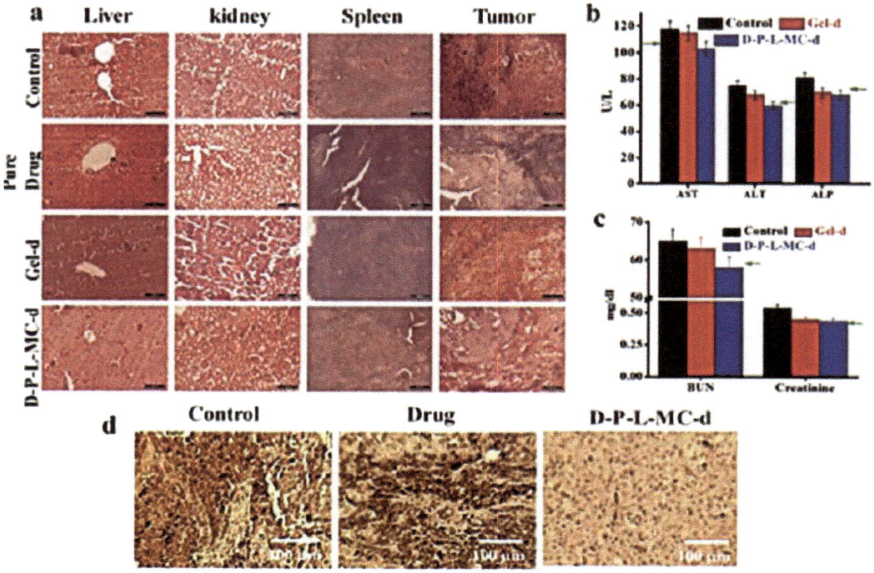

FIGURE 1.2 Effect of *in vivo* controlled drug release and its underlying effects on organs and melanoma tumor.

Note: (a) Histopathological analysis of important organs; (b) Liver function tests by estimating the serum levels of AST, ALT, and ALP across treatment groups; (c) Renal function test by measuring the serum level of BUN and creatinine across the various treatment groups; and (d) Immunohistochemistry of tumor tissues at the end of treatment regimen using melanoma inhibitory activity (MIA) antibody.

Source: Reprinted from Ref.[64]. http://creativecommons.org/licenses/by/4.0/

Another important application appeared in the literature for the medical device of cotton gauze. The cotton gauzes were grafted with a hydrogel composed of maltodextrin and poly(acrylic acid) using N-maleyl chitosan as a crosslinker. These materials exhibited antibacterial and antimicrobial properties. The loading efficiency with vancomycin (an antibiotic medicine) of hydrogel-modified cotton gauzes was increased four times in comparison to the pure ones, showing a grafting percentage-dependent manner [66]. An injectable iodine-loaded starch-g-poly(N-isopropyl acrylamide) nanogel with responsive properties was designed and developed for cancer photothermal applications. The size of the prepared hydrogel was found to be 50 nm with a spherical morphology. The intratumorally administered injection showed significant inhibition of the tumor growth. Another important finding was the anti-infection properties of Staphylococcus aureus after the photothermal therapy due to the release of iodine [67].

In the same context, the diblock poly(ethylene glycol) methyl ether (mPEG)-poly(ε-caprolactone) copolymers were grafted onto chitosan for anticancer therapy due to the pH-responsive sol-gel phase transitions because precipitation of the hydrated network at neutral pH for all temperatures was observed. Their biocompatibility and pH-responsive properties, especially at the low pH of the tumor environment, made them candidates for injectable therapy. For this reason, doxorubicin and curcumin were encapsulated in these hydrogels, showing a burst initial release followed by a delayed release for a day. The amount of curcumin released was higher than doxorubicin's [68]. Chitosan-graft-poly(ε-caprolactone) was prepared and self-assembled into micellar morphologies with sizes below 100 nm. The cytotoxicity profile of the pure nanomicelles in lung and breast cancer cell lines showed biocompatibility and improved internalization, leading to their effectiveness as DDSs. The anticancer medicine 5-fluorouracil was encapsulated into micelles, and the drug release kinetics exhibited different cumulative ratios while profiles were strongly affected by the degradation, the hydrophobic/hydrophilic ratio, and the affinity between API and the carrier [69].

1.5 DELIVERY OF PROTEINS

Polysaccharide graft copolymers could be used not only to protect and deliver molecules of small molecular weight, such as APIs but macromolecules as well. To this end, a nanogel, which consists of chondroitin sulfate conjugated with poly(L-lactide), was formulated to deliver nisin. Preformulation studies were conducted to select the optimum L-lactide content and copolymer's concentration level. *In vitro* experiments demonstrated the stimuli-responsiveness of the formulation in terms of pH and temperature fluctuations with an outstanding release profile (95%) at body temperature and acidic conditions. Besides, the nanovector was characterized by biocompatibility, a round shape, negative charge, and size equal to 180 nm at body temperature. The antimicrobial properties were examined and confirmed in comparison to the free peptide and unloaded vector [70].

A chitosan copolymer modified with deoxycholic acid and polyacrylic acid was used to create a platform for growth hormone delivery. *In vitro* and in silico techniques were utilized to deeply understand the peptide/copolymer interactions and, thus, the system's potential efficacy. The hormone/copolymer self-assembly was thermodynamically favored and characterized mainly by hydrophobic and ionic interactions. More importantly, the maintenance of the three-dimensional configuration of the protein – and thus its functionality – into the complex was demonstrated [71]. Vasquez et al. [72] studied the impact of percent amine groups on the behavior of chitosan-based nanosystems grafted with PEG moieties. Three levels were investigated, which led to different physicochemical characteristics and release properties. However, all formulations exhibited similar morphology despite the different R_g/R_H values and biocompatibility *in vitro*. The entrapment capability and the kinetics of albumin showed fluctuations between the three nanoplatforms; the difference was more obvious in acidic pH. In conclusion, varying percent amine groups could fine-tune the characteristics of the nanodelivery platform according to intended needs. Malviya et al. [73] composed novel formulations to deliver peptides through the skin. Albumin was used as a model protein and a copolymer of neem gum grafted with acrylamide as the biomaterial of choice, according to preformulation studies. Films were prepared in different copolymer concentrations, and their mechanical characteristics were interpreted. The protein was loaded completely in the films and effectively passed the skin barrier. Finally, the release kinetics of albumin were studied *in vitro*, and an extended release was declared.

Heparin was modified with PCLA via a Michael-addition reaction to create a thermoresponsive copolymer for hydrogel formation and protein loading. Lysozyme was utilized for this purpose. The amount of PCLA in the copolymer could alter the properties of the formulation, while the preferred one for further investigation was at a 5:1 PCLA: heparin ratio. The copolymer was studied for its toxicity *in vitro* and proved to be safe for use. *In vivo* experiments indicate that the gel was successfully formed in situ, and an extended release of the peptide was observed compared to other systems [74]. Another example of heparin-based nanocarriers deals with the formation of nanogels composed of polysaccharides grafted with poloxamer and a ligand to target cancer cells. The utilized protein was RNase A and was successfully loaded (80%). The particle size was around 80 nm. The formulation is also characterized by redox-responsiveness, and it is proven to release its cargo by appropriate stimulus. The addition of the ligand resulted in better cell internalization. Both properties improved the efficacy of the nano vector [75].

Furthermore, there are plenty of scientific works that add value to antidiabetic therapy, especially in oral administration. Namely, Song et al. [76] formed a PLGA-grafted dextran nanovector for the biodistribution of exenatide cargo. The copolymer was also comprised of a peptide that enables selectivity for intestinal cells. The nanocarrier is almost 140 nm in size and has a neutral charge, while its ability for peptide entrapment is around 70%. Upon extensive *in vitro* and *in vivo* experiments, the researchers concluded that the formulation is efficient for targeted therapy, could protect the peptide in physiological conditions, and has good perspectives in diabetes therapy since the therapeutic effect could last for a prolonged period. Many studies focus on insulin delivery via PEGylated chitosan either orally or intranasally. A recent example comes from Liu and colleagues, who investigated the effectiveness of different insulin-loaded nanocarriers *in vitro* and *in vivo*. In particular, they prepared chitosan-grafted copolymers with monomethyl ether PEG in various ratios of PEG moiety so as to find out which is the optimal one. Subsequently, other nanoplatforms were developed using copolymers with glyceryl monocaprylate incorporated into chitosan, with or without PEGylation. Although PEG density did not alter the physicochemical properties of the nanocarriers, it remains a crucial design parameter, influencing hydrophilicity, stability, pharmacokinetics (PK), permeation ability, and antidiabetic efficacy per oral administration [77]. Comprehensive studies on chitosan-g-PEG nanoparticles, administered intranasally, explore design factors impacting insulin delivery. The copolymers were crosslinked with tripolyphosphate (TPP) ions in varying ratios to form nanoparticulate systems, all of which were cationic with an average size of 200 nm. The mass of chitosan and PEG, along with PEG concentration in the copolymer, significantly affected the physicochemical properties of the nanocarrier, as well as *in vitro* and *in vivo* PK profiles. Ultimately, the nanoplatforms demonstrated efficient penetration through nasal mucus and satisfactory efficacy [78].

1.6 DELIVERY OF NUCLEIC ACIDS

Albeit their extensive usage for drug delivery, polysaccharide graft copolymers are utilized in gene silencing as well. Ping et al. [79] formulated a variety of polysaccharide graft copolymers via atom transfer radical polymerization comprised of chitosan and poly ((2-dimethyl amino) ethyl methacrylate), differentiating the length of the polycation segment. Complexes of the copolymer with a model plasmid DNA encoding luciferase were prepared by mixing and incubating at ambient temperature. Their properties were investigated by a gamut of techniques, and the authors concluded that the formed polyplexes with an optimal charge ratio of 6 were deemed as less toxic than other polymeric gene vectors with an appropriate size of 100–200 nm, positive zeta potential, robustness, and a good transfection ability.

Dextran and pullulan-based copolymers were formed via a two-step synthetic route: polyethyleneimine (PEI) moiety was first grafted on the polysaccharides and then DEAEM segment at two different concentrations. The polyplexes consisting of the copolymers and DNA of the p53 protein were examined thoroughly by different assays, and the ratio 5:1 proved to be the ideal one. Moreover, the 10:0.005 weight ratio of polysaccharide: DEAEM exhibited the best performance – above 90% efficiency, as a gene delivery system irrespective of the type of polysaccharide. Both these round-shaped nanoplatforms are positively charged, biocompatible, and selective to tumor cell types. Finally, *in vivo* experiments advocate good PKs [80].

Ren et al. [81] investigated chitosan derivatives for gene delivery and compared their properties. P(PEGMA-co-DMAEMA) was grafted onto a modified polysaccharide by radical coupling reaction, and the copolymer was amalgamated with luciferase plasmid DNA for complex formation. Among the polyplexes, the one with a copolymer/nucleic acid ratio of 20 incorporating a chitosan molecule of high molecular weight and folic acid as a ligand was highly distinguished for its stability as well as its transfection capability and minor cytotoxicity *in vitro*. Its size is close to 100 nm with a narrow size distribution and positive zeta potential. An innovative chitosan-based carrier was presented by Park and colleagues [82] for targeted cancer therapy. The gene vector consisted of an LMW chitosan, PEI polymer, and gambogic acid as a ligand. This polysaccharide graft copolymer displayed a good affinity to plasmid DNA encoding BCL2 nucleic acid, *in vitro* and *in vivo* efficacy, and selectivity to specific cell types in comparison with other polyplexes. *In vivo*, experiments indicate not only targeted but also sustained release.

Another gene vector for targeted treatment was introduced by Gao et al. [83] a poly(alkyl methacrylate-co-methacrylic acid) copolymer was synthesized, and its pH-sensitivity was demonstrated before it was grafted on β-glucan polysaccharide. Afterward, the copolymer was mixed with chitosan and siRNA. The complex was examined for its toxicity, ease of penetrating cells, *in vitro* and *in vivo* performance, and especially for its ability to be mainly located in inflammatory regions. The chosen formulation had good physicochemical characteristics and an excellent capability to encapsulate the nucleic acid (more than 90%). Overall, the results were encouraging, and the effectiveness of the system was also confirmed by utilizing two different siRNAs for GAPDH and TGF-β inhibition. A copolymer of PEI conjugated to rice bran polysaccharide was formed and further modified with a peptide capable of improving cell internalization and gene delivery. Complexes with a model pDNA in different charge ratios were prepared by incubation in a cell culture medium. Polyplexes were correlated *in vitro* with other polymeric vectors, including those that were marketed and available. Even though the polysaccharide-based delivery system is safe, protective to the macromolecule, and efficient at a charge ratio of 1.5, in fact, its ability to transfect cells is lower than that of the marketed available carrier (approximately 40%) [84].

Additionally, Lin and Lee compared HA and chondroitin sulfate for their affinity as gene carriers to CD44 receptors located at specific tumor cells. For this purpose, they added PLGA-PEG copolymer to the biopolymers and composed two different supramolecular complexes of polysaccharide graft copolymers and a lipoplex. Chondroitin sulfate-based polyplex was distinguished for its gene delivery properties *in vitro* assays, whereas the nano platform was characterized by an average size of 200 nm, negative charge, biocompatibility, and stability in time. The PKs of the nanocarrier depended on the pH medium, and the model protein was expressed in a discriminating way [85].

A promising hydrogel comprised of alginate and P(NIPAAm) in a 1:9 ratio was formulated by Chalanqui et al. [86] with the aim to be utilized as a gene vector in resistant cancer treatment. The thermosensitive polysaccharide graft copolymer was produced by free radical polymerization. In preformulation studies, they examined the alginate characteristics that mostly affect the hydrogel properties. The complex included a peptide to condense the plasmid DNA of miR34a RNA, while the optimal formulation exhibited modified release capability, *in vitro* non-toxicity, and the ability to protect the nucleic acid.

An interesting example of pullulan-based graft copolymers comes from the addition of ethanolamine-modified PGMA moieties via atom transfer radical polymerization to the polysaccharide. Different formulations were studied for their properties as gene delivery systems and were correlated to PEI polyplexes for their toxicity. These low toxicity complexes displayed sensitivity to reduction stimulus and a preference to be inserted into liver cells. Their effectiveness was explored *in vitro* with two different prototypes of plasmid DNA [87].

Recently, Ma and colleagues [88] composed spherical gene nanocarriers consisting of plant-based polysaccharides and PEI copolymers. The macromolecule of choice was an RNA encoding the suppression of a ligand associated with the avoidance of the immune system by cancer cells. According to preformulation studies, the optimal nanoparticle: nucleic acid ratio was 18:1, while the size of the positively charged polyplexes was close to 200 nm. The results demonstrate the nanoplatform's potential use in cancer therapy and prevention of metastatic incidents; in fact, the effectiveness of the vector is remarkable, and the gene silencing approximates 70–80% *in vitro* and *in vivo*.

Furthermore, the simultaneous loading of API and genes in nanoplatforms is frequently preferred in pharmaceutical technology, especially in mono-therapy, with the intention of synergistic effect and, thus, a more efficient and safer treatment. Jia et al. [89] formulated a smart nanocarrier for this purpose in different charge ratios. Namely, they mixed chitosan grafted copolymers of FA-PEG and PEI, silica nanosystems incorporating the anticancer drug, and an RNA encoding P-glycoprotein suppression. *In vitro* experiments showed that the well-tolerated complexes have targeting ability and stimuli-responsive properties in different reductive and pH conditions. Additionally, in charge ratios of 20:1 and 30:1, they achieved protein knockdown in a range of 30–60% and increased tumor cells' response to the therapy.

Another multifunctional delivery system was synthesized by a novel copolymer comprised of grafted pullulan in a two-step process. The chosen moieties were desoxycholic acid and PEI. Firstly, the nano platform was studied separately for its ability to deliver the p53 gene and an anticancer drug. Indeed, the physicochemical characterization suggested cationic spherical micelles of 160 nm diameter. The *in vitro* assays confirmed their biosafety as well as their ability to encapsulate the API (approximately 80% drug entrapment) and to release it mainly in an acidic environment. The optimal charge ratio for gene delivery proved to be 10. Afterward, *in vitro* and *in vivo* experiments were conducted, and the delivery of both cargos was correlated to separate treatments. The results were encouraging (80% tumor reduction), and combined therapy was established as the predominant method [90].

1.7 CHALLENGES AND LIMITATIONS

Polysaccharide graft copolymers exhibit a lot of advantages and are candidates for modified release in drug and gene delivery. However, there are still limitations and concerns about their usage that do not permit their wide clinical application. Biopolymers' natural origin complicates their yield due to batch-to-batch variability, while their chemical synthesis is difficult.

Namely, qualitative and quantitative differences depending on the source of the polysaccharide could lead to different inherent properties reflecting on the nanoparticle critical quality attributes such as their size or the encapsulation efficiency, and thus their *in vivo* behavior. From this perspective, the thermodynamic evaluation of biomaterials and their interactions is crucial [2, 91, 92]. In addition, attention must be paid to the sites of polysaccharides that are modified so as not to hide functional regions of the biopolymers and thus diminish their versatile properties. Additional techniques could contribute to confirming the maintenance of their intrinsic features. For instance, the in-depth investigation of their physicochemical and physiological characteristics moves in this direction [92]. Besides, the physicochemical and morphological characterization of the nanoassemblies by DLS, microscopy techniques, and other relevant methods is critical for the stability assessment of the systems. It is well-documented that colloidal suspensions are thermodynamically unstable systems [93]. As a result, research works should focus on the storage stability of the formulations in terms of size and shape, while their *in vivo* stability should be examined, and their functionality should be ensured in serum conditions.

Considering the synthetic routes, there are still problems that result in biomaterials with uncontrollable characteristics and low purity. It is widely known that polymerization methods should become more robust and effective [94, 95]. Another issue that the scientific community deals with is organic solvent usage and environmental repercussions relevant to polymerization techniques. There are still not many efficient alternative methods, though [2, 95, 96]. Furthermore, intrinsic features of copolymers, such as lower critical solution temperature (LCST), cannot be easily correlated among other studies or *in vivo* performance due to a lack of information and different ways of determination. In this manner, the reproducibility of results, as well as fast clinical translation, remains difficult to achieve [1].

Moreover, nanoparticles show inherent toxicity due mainly to their size but, in some cases, to their PK profile as well. Although most systems that are presented in this chapter are deemed biocompatible, it is vital for nanotoxicity and immunogenicity assays to be conducted *in vitro* and *in vivo* so that the biosafety can be confirmed. The toxicological aspects, not only of the formulation but also of the biomaterials that it consists of, should also be considered and correlated to the preferred route of administration [97]. Finally, there is not much research focusing on *in vivo* experiments to confirm *in vitro* results and prove the effectiveness of the biomaterials [2].

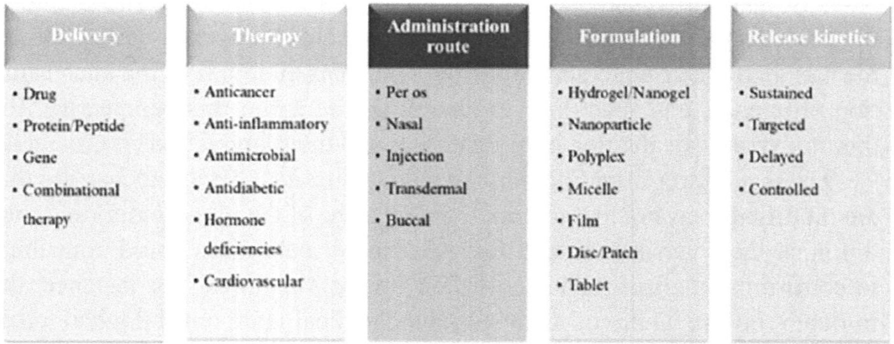

FIGURE 1.3 Polysaccharide graft copolymer formulations and their pharmaceutical applications.

1.8 CONCLUSIONS

This chapter deals with the synthesis, design, and development of polysaccharide graft-copolymers for drug delivery applications (Figure 1.3). Different nanosystems that are self-assembled from polysaccharide graft copolymers were presented in terms of their physicochemical, morphological, and biological characteristics, as well as their ability to encapsulate LMW molecules of pharmaceutical interest, as well as nucleic acids and proteins/peptides. Polysaccharide graft copolymer formulations are now in preclinical evaluation, mainly for anticancer therapy and for hormone deficiencies and cardiovascular diseases. Figure 1.3 summarizes the polysaccharide graft copolymer formulations and their pharmaceutical applications. They are utilized via the majority of administration routes and exhibit all the release kinetic profiles of the encapsulated APIs. Considering the increasing number of studies in recent years, polysaccharide graft copolymers are "intelligent" biomaterials that will be used very soon as platforms for the development of nanomedicines with added value for patients.

KEYWORDS

- **cellulose sodium salts**
- **dynamic light scattering**
- **electrophoretic light scattering**
- **poly (lactic-co-glycolic acid)**
- **quaternized cyclodextrin**

REFERENCES

1. Marsili, L., Dal Bo, M., Berti, F., & Toffoli, G. (2021). Chitosan-based biocompatible copolymers for thermoresponsive drug delivery systems: On the development of a standardization system. *Pharmaceutics, 13*(11), 1876.
2. Torres, F. G., & De-la-Torre, G. E. (2022). Synthesis, characteristics, and applications of modified starch nanoparticles: A review. *International Journal of Biological Macromolecules, 194*, 289–305.
3. Işıklan, N., & Tokmak, Ş. (2018). Microwave based synthesis and spectral characterization of thermo-sensitive poly (N, N-diethylacrylamide) grafted pectin copolymer. *International Journal of Biological Macromolecules, 113*, 669–680.
4. Carvalho, A. M., Valcarcel, J., Soares da Costa, D., Gomes, M., Vázquez, J. A., Reis, R. L., et al. (2022). Hyaluronan brush-like copolymers promote CD44 decluttering in breast cancer cells. *ACS Applied Materials & Interfaces, 14*(37), 41779–41789.
5. Kumar, D., Gihar, S., Shrivash, M. K., Kumar, P., & Kundu, P. P. (2020). A review on the synthesis of graft copolymers of chitosan and their potential applications. *International Journal of Biological Macromolecules, 163*, 2097–2112.
6. Jha, S., Malviya, R., Fuloria, S., Sundram, S., Subramaniyan, V., Sekar, M., et al. (2022). Characterization of microwave-controlled polyacrylamide graft copolymer of tamarind seed polysaccharide. *Polymers, 14*(5), 1037.
7. Zhang, Y., Sun, T., & Jiang, C. (2018). Biomacromolecules as carriers in drug delivery and tissue engineering. *Acta Pharmaceutica Sinica B, 8*(1), 34–50.
8. Six, J. L., & Ferji, K. (2019). Polymerization induced self-assembly: An opportunity toward the self-assembly of polysaccharide-containing copolymers into high-order morphologies. *Polymer Chemistry, 10*(1), 45–53.
9. Garcia-Valdez, O., Champagne, P., & Cunningham, M. F. (2018). Graft modification of natural polysaccharides via reversible deactivation radical polymerization. *Progress in Polymer Science, 76*, 151–173.
10. Atanase, L. I., Desbrieres, J., & Riess, G. (2017). Micellization of synthetic and polysaccharides-based graft copolymers in aqueous media. *Progress in Polymer Science, 73*, 32–60.
11. Bhattacharya, A., & Misra, B. N. (2004). Grafting: A versatile means to modify polymers: Techniques, factors and applications. *Progress in Polymer Science, 29*(8), 767–814.
12. Uhrig, D., & Mays, J. (2011). Synthesis of well-defined multigraft copolymers. *Polymer Chemistry, 2*(1), 69–76.
13. Mittal, H., Ray, S. S., & Okamoto, M. (2016). Recent progress on the design and applications of polysaccharide-based graft copolymer hydrogels as adsorbents for wastewater purification. *Macromolecular Materials and Engineering, 301*(5), 496–522.
14. Basu, A., Kunduru, K. R., Abtew, E., & Domb, A. J. (2015). Polysaccharide-based conjugates for biomedical applications. *Bioconjugate Chemistry, 26*(8), 1396–1412.
15. Guo, R., & McGrath, J. E. (2012). Aromatic polyethers, polyetherketones, polysulfides, and polysulfones. In K. Matyjaszewski & M. Möller (Eds.), *Polymer Science: A Comprehensive Reference* (pp. 377–430). Elsevier.
16. Youssouf, L., Bhaw-Luximon, A., Diotel, N., Catan, A., Giraud, P., & Gimié, F., et al. (2019). Enhanced effects of curcumin encapsulated in polycaprolactone-grafted

oligocarrageenan nano micelles, a novel nanoparticle drug delivery system. *Carbohydrate Polymers, 217*, 35–45.
17. Zuppolini, S., Maya, I. C., Diodato, L., Guarino, V., Borriello, A., & Ambrosio, L. (2020). Self-associating cellulose-graft-poly(ε-caprolactone) to design nanoparticles for drug release. *Materials Science and Engineering: C, 108*, 110385.
18. Abbasi, S., Yousefi, G., Firuzi, O., & Mohammadi-Samani, S. (2016). Design and cell cytotoxicity assessment of palmitoylated polyethylene glycol-grafted chitosan as nano micelle carrier for paclitaxel. *Journal of Applied Polymer Science, 133*(13).
19. Carvalho, L. T., Moraes, R. M., Teixeira, A. J. R. M., Tada, D. B., Alves, G. M., Lacerda, T. M., et al. (2021). Development of pullulan-based carriers for controlled release of hydrophobic ingredients. *Journal of Applied Polymer Science, 138*(45), 51344.
20. Nishimura, T., Shishi, S., Sasaki, Y., & Akiyoshi, K. (2020). Thermoresponsive polysaccharide graft polymer vesicles with tunable size and structural memory. *Journal of the American Chemical Society, 142*(27), 11784–11790.
21. Deng, J. R., Zhao, C. L., & Wu, Y. X. (2020). Antibacterial and pH-responsive quaternized hydroxypropyl cellulose-g-poly (THF-co-epichlorohydrin) graft copolymer: Synthesis, characterization and properties. *Chinese Journal of Polymer Science, 38*(7), 704–714.
22. Gao, Y. Z., Chen, J. C., Cui, Z., Zhao, C. L., & Wu, Y. X. (2022). Biocompatible propylene glycol alginate-g-polytetrahydrofuran amphiphilic graft copolymers for highly effective drug carriers. *Polymer, 246*, 124706.
23. Zhao, C. L., Gao, Y. Z., Wu, M. Y., Zhang, H. T., & Wu, Y. X. (2021). Biocompatible, hemocompatible and antibacterial acylated dextran-g-polyisobutylene graft copolymers with silver nanoparticles. *Chinese Journal of Polymer Science, 39*(12), 1550–1561.
24. Lavikainen, J., Dauletbekova, M., Toleutay, G., Kaliva, M., Chatzinikolaidou, M., Kudaibergenov, S. E., et al. (2021). Poly(2-ethyl-2-oxazoline) grafted gellan gum for potential application in transmucosal drug delivery. *Polymer Advanced Technologies, 32*(7), 2770–2780.
25. Opiyo, G., & Jin, J. (2021). Recent progress in switchable RAFT agents: Design, synthesis and application. *European Polymer Journal, 159*, 110713.
26. Ikkene, D., Arteni, A. A., Ouldali, M., Francius, G., Brûlet, A., Six, J. L., et al. (2021). Direct access to polysaccharide-based vesicles with a tunable membrane thickness in a large concentration window via polymerization-induced self-assembly. *Biomacromolecules, 22*(7), 3128–3137.
27. Ikkene, D., Arteni, A. A., Song, H., Laroui, H., Six, J. L., & Ferji, K. (2020). Synthesis of dextran-based chain transfer agent for RAFT-mediated polymerization and glyco-nanoobjects formulation. *Carbohydrate Polymers, 234*, 115943.
28. Hatton, F. L., Ruda, M., Lansalot, M., D'Agosto, F., Malmström, E., & Carlmark, A. (2016). Xyloglucan-functional latex particles via RAFT-mediated emulsion polymerization for the biomimetic modification of cellulose. *Biomacromolecules, 17*(4), 1414–1424.
29. Ferji, K., Venturini, P., Cleymand, F., Chassenieux, C., & Six, J. L. (2018). *In situ* glyco-nanostructure formulation via photo-polymerization-induced self-assembly. *Polymer Chemistry, 9*(21), 2868–2872.
30. Pilipenko, I. M., Korzhikov-Vlakh, V. A., Zakharova, N. V., Urtti, A., & Tennikova, T. B. (2020). Thermo- and pH-sensitive glycosaminoglycans derivatives obtained by controlled grafting of poly(N-isopropylacrylamide). *Carbohydrate Polymers, 248*, 116764.

31. Lorandi, F., Fantin, M., & Matyjaszewski, K. (2022). Atom transfer radical polymerization: A mechanistic perspective. *Journal of the American Chemical Society, 144*(34), 15413–15430.
32. Yuan, W., Zou, H., & Shen, J. (2016). Amphiphilic graft copolymers with ethyl cellulose backbone: Synthesis, self-assembly and tunable temperature–CO2 response. *Carbohydrate Polymers, 136*, 216–223.
33. Ferji, K., Nouvel, C., Babin, J., Li, M. H., Gaillard, C., Nicol, E., et al. (2015). Polymersomes from amphiphilic glycopolymers containing polymeric liquid crystal grafts. *ACS Macro Letters, 4*(10), 1119–1122.
34. Ribas-Massonis, A., Cicujano, M., Duran, J., Besalú, E., & Poater, A. (2022). Free-radical photopolymerization for curing products for refinish coatings market. *Polymers, 14*(14), 2856.
35. Wen, Y., & Oh, J. K. (2015). Intracellular delivery cellulose-based bionanogels with dual temperature/pH-response for cancer therapy. *Colloids and Surfaces B: Biointerfaces, 133*, 246–253.
36. Ortiz, J. A., Matsuhiro, B., Zapata, P. A., Corrales, T., & Catalina, F. (2018). Preparation and characterization of maleoylagarose/PNIPAAm graft copolymers and formation of polyelectrolyte complexes with chitosan. *Carbohydrate Polymers, 182*, 81–91.
37. Liu, M., Wen, Y., Song, X., Zhu, J. L., & Li, J. (2019). A smart thermoresponsive adsorption system for efficient copper ion removal based on alginate-g-poly(N-isopropylacrylamide) graft copolymer. *Carbohydrate Polymers, 219*, 280–289.
38. Deng, Z., & Liu, S. (2020). Emerging trends in solution self-assembly of block copolymers. *Polymer, 207*, 122914.
39. Pinto-Gómez, C., Pérez-Murano, F., Bausells, J., Villanueva, L. G., & Fernández-Regúlez, M. (2020). Directed self-assembly of block copolymers for the fabrication of functional devices. *Polymers, 12*(10), 2432.
40. Hemmati, K., & Ghaemy, M. (2016). Synthesis of new thermo/pH-sensitive drug delivery systems based on tragacanth gum polysaccharide. *International Journal of Biological Macromolecules, 87*, 415–425.
41. Yang, Y., Guo, T., Xu, J., Xiong, Y., Cui, X., Ke, Y., et al. (2021). Micelle nano vehicles for co-delivery of *Lepidium meyenii* Walp. (maca) polysaccharide and chloroquine to tumor-associated macrophages for synergistic cancer immunotherapy. *International Journal of Biological Macromolecules, 189*, 577–589.
42. Pranantyo, D., Kang, E. T., & Chan-Park, M. B. (2021). Smart nano micelles with bacterial infection-responsive disassembly for selective antimicrobial applications. *Biomaterials Science, 9*(5), 1627–1638.
43. Sakulwech, S., Lourith, N., Kanlayavattanakul, M., Phunpee, S., Suktham, K., Surassmo, S., et al. (2022). Nanocomplex of quaternized cyclodextrin grafted chitosan and hyaluronic acid for a skin delivery. *Colloids and Surfaces B: Bio Interfaces, 220*, 112920.
44. Shutava, T. G., Livanovich, K. S., & Pankov, V. V. (2018). Synergetic effect of polyethylene glycol-grafted chitosan and bovine serum albumin on colloidal stability of polyelectrolyte nano capsules. *Colloids and Surfaces A: Physicochemical and Engineering Aspects, 539*, 69–79.
45. Raorane, C. J., Shastri, D., Parveen, A. S., Haldhar, R., Raj, V., & Kim, S. C. (2022). Grafted chitosan-hyaluronic acid (CS-g-poly (MA-co-AN) HA) complex inhibits fluconazole-resistant *Candida albicans* biofilm formation. *Antibiotics, 11*(7), 950.

46. Chai, M., Gao, Y., Liu, J., Deng, Y., Hu, D., Jin, Q., et al. (2020). Polymyxin B-polysaccharide polyion nanocomplex with improved biocompatibility and unaffected antibacterial activity for acute lung infection management. *Advanced Healthcare Materials, 9*(3), 1901542.
47. Ma, C., Duan, Y., Wu, C., Meng, E., Li, P., Zhang, Z., et al. (2021). Spatiotemporally co-delivery of triple therapeutic drugs via HA-coating nano systems for enhanced immunotherapy. *Asian Journal of Pharmaceutical Sciences, 16*(5), 653–664.
48. Luckanagul, J. A., Ratnatilaka Na Bhuket, P., Muangnoi, C., Rojsitthisak, P., Wang, Q., & Rojsitthisak, P. (2021). Self-assembled thermoresponsive nanogel from grafted hyaluronic acid as a biocompatible delivery platform for curcumin with enhanced drug loading and biological activities. *Polymers, 13*(2), 276.
49. Klein, J. M., de Lima, V. S., Haack, M., da Feira, J. M. C., & de Camargo Forte, M. M. (2019). Amino-acid-modified cashew gum flocculant: Synthesis, characterization, and application. *Journal of Polymers and the Environment, 27*(7), 1465–1474.
50. Qu, J., Meng, Q., Lin, X., Han, W., Jiang, Q., Wang, L., et al. (2021). Microwave-assisted synthesis of β-cyclodextrin functionalized celluloses for enhanced removal of Pb (II) from water: Adsorptive performance and mechanism exploration. *Science of the Total Environment, 752*, 141854.
51. Marques, N. d. N., Balaban, R. d. C., Halila, S., & Borsali, R. (2018). Synthesis and characterization of carboxymethylcellulose grafted with thermoresponsive side chains of high LCST: The high temperature and high salinity self-assembly dependence. *Carbohydrate Polymers, 184*, 108–117.
52. Ikkene, D., Arteni, A. A., Ouldali, M., Six, J. L., & Ferji, K. (2020). Self-assembly of amphiphilic copolymers containing polysaccharide: PISA versus nanoprecipitation, and the temperature effect. *Polymer Chemistry, 11*(29), 4729–4740.
53. Zoratto, N., Grillo, I., Matricardi, P., & Dreiss, C. A. (2019). Supramolecular gels of cholesterol-modified gellan gum with disc-like and worm-like micelles. *Journal of Colloid and Interface Science, 556*, 301–312.
54. Yiğitoğlu, M., Aydın, G., & Işıklan, N. (2014). Microwave-assisted synthesis of alginate-g-polyvinylpyrrolidone copolymer and its application in controlled drug release. *Polymer Bulletin, 71*(2), 385–414.
55. Kaity, S., Isaac, J., Kumar, P. M., Bose, A., Wong, T. W., & Ghosh, A. (2013). Microwave-assisted synthesis of acrylamide grafted locust bean gum and its application in drug delivery. *Carbohydrate Polymers, 98*(1), 1083–1094.
56. Usha Rani, G., Konreddy, A. K., Mishra, S., & Sen, G. (2014). Synthesis and applications of polyacrylamide grafted agar as a matrix for controlled drug release of 5-ASA. *International Journal of Biological Macromolecules, 65*, 375–382.
57. Delorme, V., Lichon, L., Mahindad, H., Hunger, S., Laroui, N., & Daurat, M., et al. (2020). Reverse poly(ε-caprolactone)-g-dextran graft copolymers: Nanocarriers for intracellular uptake of anticancer drugs. *Carbohydrate Polymers, 232*, 115764.
58. Carneiro, M. J. M., Paula, C. B. A., Ribeiro, I. S., de Lima, L. R. M., Ribeiro, F. O. S., Silva, D. A., et al. (2021). Dual responsive dextran-graft-poly (N-isopropylacrylamide)/doxorubicin prodrug via Schiff base reaction. *International Journal of Biological Macromolecules, 185*, 390–402.
59. Ribeiro, I. S., Pontes, F. J. G., Carneiro, M. J. M., Sousa, N. A., Pinto, V. P. T., Ribeiro, F. O. S., et al. (2021). Poly(ε-caprolactone) grafted cashew gum nanoparticles as an

epirubicin delivery system. *International Journal of Biological Macromolecules, 179*, 314–323.

60. Abhishek, Rimpy, & Ahuja, M. (2018). *Moringa* gum-g-poly(N-vinyl-2-pyrrolidone)–A potential buccoadhesive polymer. *International Journal of Biological Macromolecules, 109*, 732–739.
61. Gao, Y. Z., Chang, T. X., & Wu, Y. X. (2019). *In-situ* synthesis of acylated sodium alginate-g-(tetrahydrofuran5-b-polyisobutylene) terpolymer/Ag-NPs nanocomposites. *Carbohydrate Polymers, 219*, 201–209.
62. Zhao, C. L., Deng, J. R., Gao, Y. Z., & Wu, Y. X. (2021). Hemocompatible, biocompatible and antifouling acylated dextran-g-polytetrahydrofuran graft copolymer with silver nanoparticles: Synthesis, characterization and properties. *Materials Science and Engineering: C, 123*, 111998.
63. Bordallo, E., Torneiro, M., & Lazzari, M. (2020). Dissolution of amorphous nifedipine from micelle-forming carboxymethylcellulose derivatives. *Carbohydrate Polymers, 247*, 116699.
64. Shukla, A., Singh, A. P., & Maiti, P. (2021). Injectable hydrogels of newly designed brush biopolymers as sustained drug-delivery vehicle for melanoma treatment. *Signal Transduction and Targeted Therapy, 6*(1), 63.
65. Chopra, L., Chohan, J. S., Sharma, S., Pelc, M., & Kawala-Sterniuk, A. (2022). Multifunctional modified chitosan biopolymers for dual applications in biomedical and industrial field: Synthesis and evaluation of thermal, chemical, morphological, structural, *in vitro* drug-release rate, swelling and metal uptake studies. *Sensors, 22*(9), 3454.
66. Meléndez-Ortiz, H. I., Betancourt-Galindo, R., Puente-Urbina, B., Sánchez-Orozco, J. L., & Ledezma, A. (2022). Antimicrobial cotton gauzes modified with poly (acrylic acid-co-maltodextrin) hydrogel using chitosan as crosslinker. *International Journal of Biological Macromolecules, 198*, 119–127.
67. Fan, M., Li, M., Wang, X., Liao, Y., Wang, H., Rao, J., et al. (2022). Injectable thermosensitive iodine-loaded starch-g-poly(N-isopropylacrylamide) hydrogel for cancer photothermal therapy and anti-infection. *Macromolecular Rapid Communications, 43*(18), 2200203.
68. Jommanee, N., Chanthad, C., & Manokruang, K. (2018). Preparation of injectable hydrogels from temperature and pH responsive grafted chitosan with tuned gelation temperature suitable for tumor acidic environment. *Carbohydrate Polymers, 198*, 486–494.
69. Gu, C., Le, V., Lang, M., & Liu, J. (2014). Preparation of polysaccharide derivatives chitosan-graft-poly(ε-caprolactone) amphiphilic copolymer micelles for 5-fluorouracil drug delivery. *Colloids and Surfaces B: Bio Interfaces, 116*, 745–750.
70. Ghaeini-Hesaroeiye, S., Boddohi, S., & Vasheghani-Farahani, E. (2020). Dual responsive chondroitin sulfate based nanogel for antimicrobial peptide delivery. *International Journal of Biological Macromolecules, 143*, 297–304.
71. Khanmohammadi, S., Mehrnejad, F., Lotfi-Sousefi, Z., Yahyaei, M., & Naderi-manesh, H. (2021). Design and synthesis of polyacrylic acid/deoxycholic acid-modified chitosan copolymer and a close inspection of human growth hormone-copolymer interactions: An experimental and computational study. *Colloids and Surfaces B: Bio Interfaces, 206*, 111956.

72. Vasquez, D., Milusheva, R., Baumann, P., Constantin, D., Chami, M., & Palivan, C. G. (2014). The amine content of PEGylated chitosan *Bombyx mori* nanoparticles acts as a trigger for protein delivery. *Langmuir, 30*(4), 965–975.
73. Malviya, R. (2020). Non-invasive drug delivery system for the delivery of protein/peptide using neem gum and its derivatives. *Bio Interface Research in Applied Chemistry, 10*(3), 5460–5465.
74. Sim, H. J., Thambi, T., & Lee, D. S. (2015). Heparin-based temperature-sensitive injectable hydrogels for protein delivery. *Journal of Materials Chemistry B, 3*(45), 8892–8901.
75. Nguyen, D. H., Joung, Y. K., Choi, J. H., Moon, H. T., & Park, K. D. (2011). Targeting ligand-functionalized and redox-sensitive heparin-Pluronic nanogels for intracellular protein delivery. *Biomedical Materials, 6*(5), 055004.
76. Song, Y., Shi, Y., Zhang, L., Hu, H., Zhang, C., Yin, M., et al. (2019). Synthesis of CSK-DEX-PLGA nanoparticles for the oral delivery of exenatide to improve its mucus penetration and intestinal absorption. *Molecular Pharmaceutics, 16*(2), 518–532.
77. Liu, C., Kou, Y., Zhang, X., Dong, W., Cheng, H., & Mao, S. (2019). Enhanced oral insulin delivery via surface hydrophilic modification of chitosan copolymer-based self-assembly polyelectrolyte nanocomplex. *International Journal of Pharmaceutics, 554*, 36–47.
78. Zhang, X., Zhang, H., Wu, Z., Wang, Z., Niu, H., & Li, C. (2008). Nasal absorption enhancement of insulin using PEG-grafted chitosan nanoparticles. *European Journal of Pharmaceutics and Biopharmaceutics, 68*(3), 526–534.
79. Ping, Y., Liu, C. D., Tang, G. P., Li, J. S., Li, J., Yang, W. T., et al. (2010). Functionalization of chitosan via atom transfer radical polymerization for gene delivery. *Advanced Functional Materials, 20*(18), 3106–3116.
80. Sherly, M. C. D., Mr, R., & V. S., H. (2020). Cationized dextran and pullulan modified with diethyl aminoethyl methacrylate for gene delivery in cancer cells. *Carbohydrate Polymers, 242*, 116426.
81. Ren, H., Liu, S., Yang, J., Zhang, X., Zhou, H., Chen, J., et al. (2016). N, N, N-trimethyl chitosan modified with well-defined multifunctional polymer modules used as pDNA delivery vector. *Carbohydrate Polymers, 137*, 222–230.
82. Park, S. C., Heo, H., & Jang, M. K. (2022). Polyethyleneimine grafted-chitosan-based Gambogic acid copolymers for targeting cancer cells overexpressing transferrin receptors. *Carbohydrate Polymers, 277*, 118755.
83. Gao, H., Wang, S., Long, Q., Cheng, R., Lian, W., Koivuniemi, A., et al. (2023). Rational design of a polysaccharide-based viral mimicry nanocomplex for potent gene silencing in inflammatory tissues. *Journal of Controlled Release, 357*, 120–132.
84. Liu, L., Yan, Y., Ni, D., Wu, S., Chen, Y., Chen, X., et al. (2020). TAT-functionalized PEI-grafting rice bran polysaccharides for safe and efficient gene delivery. *International Journal of Biological Macromolecules, 146*, 1076–1086.
85. Lin, W. L., & Lee, W. C. (2018). Polysaccharide-modified nanoparticles with intelligent CD44 receptor targeting ability for gene delivery. *International Journal of Nanomedicine, 13*, 3989–4002.
86. Chalanqui, M. J., Pentlavalli, S., McCrudden, C., Chambers, P., Ziminska, M., Dunne, N., et al. (2019). Influence of alginate backbone on efficacy of thermo-responsive alginate-g-P(NIPAAm) hydrogel as a vehicle for sustained and controlled gene delivery. *Materials Science and Engineering: C, 95*, 409–421.

87. Yang, X. C., Niu, Y. L., Zhao, N. N., Mao, C., & Xu, F. J. (2014). A biocleavable pullulan-based vector via ATRP for liver cell-targeting gene delivery. *Biomaterials, 35*(12), 3873–3884.
88. Ma, M., Liu, X., Ma, C., Guo, R., Zhang, X., Zhang, Z., et al. (2023). Enhancing the antitumor immunosurveillance of PD-L1-targeted gene therapy for metastatic melanoma using cationized Panax Noto ginseng polysaccharide. *International Journal of Biological Macromolecules, 226*, 1309–1318.
89. Jia, L., Li, Z., Zheng, D., Li, Z., & Zhao, Z. (2021). A targeted and redox/pH-responsive chitosan oligosaccharide derivatives based nanohybrids for overcoming multidrug resistance of breast cancer cells. *Carbohydrate Polymers, 251*, 117008.
90. Chen, L., Ji, F., Bao, Y., Xia, J., Guo, L., Wang, J., et al. (2017). Biocompatible cationic pullulan-g-desoxycholic acid-g-PEI micelles used to co-deliver drug and gene for cancer therapy. *Materials Science and Engineering: C, 70*, 418–429.
91. Debele, T. A., Mekuria, S. L., & Tsai, H. C. (2016). Polysaccharide based nanogels in the drug delivery system: Application as the carrier of pharmaceutical agents. *Materials Science and Engineering: C, 68*, 964–981.
92. Hu, Y., Li, Y., & Xu, F. J. (2017). Versatile functionalization of polysaccharides via polymer grafts: From design to biomedical applications. *Accounts of Chemical Research, 50*(2), 281–292.
93. Wang, Y., Zheng, Y., Zhang, L., Wang, Q., & Zhang, D. (2013). Stability of nanosuspensions in drug delivery. *Journal of Controlled Release, 172*(3), 1126–1141.
94. Bhavsar, C., Momin, M., Gharat, S., & Omri, A. (2017). Functionalized and graft copolymers of chitosan and its pharmaceutical applications. *Expert Opinion on Drug Delivery, 14*(10), 1189–1204.
95. Maji, B., & Maiti, S. (2021). Chemical modification of xanthan gum through graft copolymerization: Tailored properties and potential applications in drug delivery and wastewater treatment. *Carbohydrate Polymers, 251*, 117095.
96. Sarder, R., Piner, E., Rios, D. C., Chacon, L., Artner, M. A., Barrios, N., et al. (2022). Copolymers of starch, a sustainable template for biomedical applications: A review. *Carbohydrate Polymers, 278*, 118973.
97. Patnaik, S., Gorain, B., Padhi, S., Choudhury, H., Gabr, G. A., Md, S., et al. (2021). Recent update of toxicity aspects of nanoparticulate systems for drug delivery. *European Journal of Pharmaceutics and Biopharmaceutics, 161*, 100–119.

CHAPTER 2

Innovative Drug Delivery Systems Using Grafted Copolymers: Recent Advances and Challenges

JIGAR VYAS,[1] NENSI RAYTTHATHA,[1] and BHUPENDRA PRAJAPATI[2]

[1]*Faculty of Pharmacy, Sigma University, Vadodara, Gujarat, India*

[2]*Shree S. K. Patel College of Pharmaceutical Education and Research, Ganpat University, Mehsana, Gujarat, India*

ABSTRACT

Grafted copolymers consist of one polymer chain grafted onto the backbone of another, resulting in a copolymer with two or more distinct polymer chains. Grafting can occur through physical adsorption or covalent bonding. These copolymers are highly attractive for various applications due to their distinctive properties, particularly in drug delivery systems (DDSs). Their ability to encapsulate pharmaceuticals, protect them from degradation, and target specific tissues or cells makes them ideal candidates for drug delivery applications. In general, grafted copolymers have demonstrated potential for creating novel materials with enhanced properties and tailored functionality. While they have already shown promise as drug delivery vehicles, further research is needed to fully explore their potential in diverse fields.

2.1 INTRODUCTION TO INNOVATIVE DRUG DELIVERY SYSTEMS (IDDS)

The discovery of novel chemical entities has long been a challenge in the history of medicine. More recently, dosage form design has been recognized as a critical factor in the success of drugs. Not only have the time

Graft Copolymers of Polysaccharides for Drug Delivery Systems.
Rishabha Malviya and Sonali Sundram (Eds.)

and costs associated with researching new compounds increased, but it has also been found that many products can be improved through integration into rate-controlled forms of administration. Conventional formulations have historically served as vehicles for introducing compounds into the body, but the limitations of these systems highlight the need for innovative drug delivery systems (DDSs) to improve patient outcomes. The term "intelligent drug delivery systems" (IDDS) refers to various approaches and technologies used to develop advanced pharmacological formulations [1]. These dosage forms are optimized for particle size, permeability, and targeted delivery to specific sites. The objective is to enhance the efficacy of biotherapeutic drugs compared to their performance in traditional dosage forms. New DDSs can target specific areas of the body, improving treatment effectiveness while minimizing side effects. Such systems can enhance drug delivery by preventing degradation, improving solubility, and increasing absorption by target cells. IDDS represents a significant area of pharmaceutical research, with the potential to improve patient outcomes and reduce healthcare costs [2].

2.1.1 *CLASSIFICATION OF IDDS*

IDDS can also be classified based on their method of administration, including intravenous, oral, inhalation, and transdermal drug delivery systems. The main factors in the selection of drug delivery methods are the physicochemical properties of drugs, intended pharmacokinetics and pharmacodynamic profiles, mode of administration, and target tissue or organ, among many others. IDDS can be classified based on types of delivery systems, as shown in Figure 2.1.

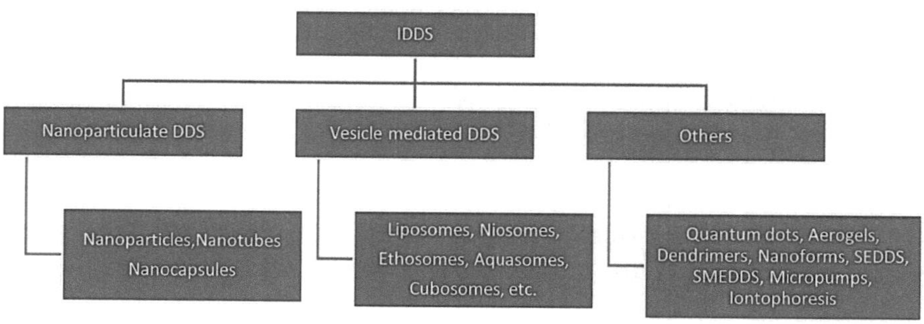

FIGURE 2.1 IDDS classification.

2.1.1.1 NANOPARTICULATE DRUG DELIVERY SYSTEM

Nanoparticulate drug delivery techniques include nanoparticles, nanotubes, and nanocapsules with a size range from 1 to 1000 nanometers, which are intended to increase therapeutic effectiveness and safety. Nanoparticles are minuscule particles built of a wide range of substances, including polymers, metals, ceramics have a broad spectrum of chemical and physical properties. Nanotubes are a form of nanoparticles consisting of a hollow tube structure and nano-diameter [3]. Nanocapsules are nanoparticles with a shell encompassing a core, wherein the shell can be constructed of a range of materials, including proteins, polymers, and lipids. In contrast, the core can have an array of payloads, including medicines, genes, or imaging agents [4].

2.1.1.2 VESICLE-MEDIATED DRUG DELIVERY SYSTEM

A vesicle-mediated DDS encapsulates pharmaceuticals or therapeutic agents within vesicles, which are tiny membrane frameworks, for controlled and targeted drug release to a specific site in the body. Natural or manufactured vesicles used in drug delivery may vary in size between nanometers to micrometers. Examples include liposomes, niosomes, aquasomes, ethosomes, and other vesicle-mediated DDSs. Liposomes are spherical vesicles comprising a lipid bilayer capable of encapsulating both hydrophobic and hydrophilic drugs. Polymers are artificial or natural polymers that may be engineered to release medicament in a controlled manner [5].

Niosomes are a vesicle-mediated system for drug delivery made up of non-ionic surfactants like Span and Tween, as well as cholesterol. They are structurally similar to liposomes, but the main distinction is that niosomes comprise non-ionic surfactants rather than phospholipids [6].

Ethosomes are lipid-based nanocarriers made up of phospholipids, alcohol, and water. Drug solubility, and stability, effectiveness can all be improved by tailoring ethosomes. They may be preloaded with hydrophilic and lipophilic drugs, and their surfaces can be tailored to target specific cells or tissues [7].

Aquasomes are vesicle-mediated DDSs comprising an inner core of biodegradable polymers or ceramic nanoparticles covered with a coating of surfactants and drug-adsorbed. The hydrophilic surfactant coating on the surface of nanoparticles allows it to spread within the fluid and be readily taken up by cells [8].

Cubosomes are lipid nanoparticles that self-assemble and comprise a lipid bilayer that produces a bicontinuous cubic phase that divides two internal water channels. Cubosomes' unusual shape makes them an appealing drug delivery vehicle since they may contain both hydrophilic and hydrophobic drugs, preserve them against decomposition, and allow for continuous release. They are also biodegradable and biocompatible, making them suitable for use [9].

2.1.1.3 OTHER DRUG DELIVERY SYSTEM

These systems are intended for sustained and targeted drug delivery, allowing for consistent drug release over an extended period of time. Quantum dots are semiconductor nanoparticles having distinct optical and electrical features such as high quantum yield, size-tunable fluorescence, and photobleaching resistance. They have been utilized to image biological samples as fluorescent probes.

Aerogels are highly porous materials with low density, large surface area, and good thermal insulation. They are employed in many applications, including catalytic supports, sensors, devices for storing energy, and insulation substances [10]. Dendrimers are monodispersed, highly branching macromolecules having a well-defined structure and size. Their applications include drug delivery, catalysis, sensing, and imaging [11].

Nanoforms are any nanoscale substances that have been purposefully created or developed, including nanowires, nanotubes, and nanoparticles. They contain unique features that allow them to be used in an array of applications [12].

Self-emulsifying drug delivery systems (SEDDS) and self-micro-emulsifying drug delivery systems (SMEDDS) are formulations made from lipids that potentially improve solubility and bioavailability of drugs. When they come into contact with watery fluids in the gastrointestinal (GI) system, they may produce tiny oil-in-water emulsions or microemulsions, improving drug absorption and therapeutic efficacy [13].

Micropumps may control the delivery of minute amounts of fluids or gases and can be employed for the administration of pharmaceuticals, microfluidics, and analytical chemistry [14]. Iontophoresis is a non-invasive drug delivery method that employs an electric field to propel charged molecules such as drugs or ions across the skin or other biological barriers. It has been utilized in the treatment of different medical problems like hyperhidrosis and musculoskeletal ailments, as well as for transdermal administration of drugs [15].

2.1.2 CONSTITUENTS OF INNOVATIVE DRUG DELIVERY SYSTEM

The main components of IDDS are Active pharmaceutical ingredients (APIs), carrier systems, surfactants, stabilizers and polymers. The purpose of the API is to deliver a therapeutic effect. The carrier system acts as a vehicle to transport the API to its intended site in the body. The carrier system is intended to preserve the drugs from degradation and elimination; to improve its stability, and to manage its release, resulting in improved effectiveness of the medicament, safety, and patient compliance. Surfactants, also known as surface-active agents, are extensively employed due to their unique features that can increase the effectiveness of systems for drug delivery [16]. Stabilizers like EDTA, trehalose, polysorbate 80, ascorbic acid, and sodium metabisulfite are key components that aid the stability of the formulation throughout storage, transit, and administration. Stabilizers can help avoid or reduce chemical, physical, or biological alterations in the DDS, ensuring it maintains effectiveness and is safe for its intended usage. Polymers are important in DDSs as they might be modified to give characteristics including biodegradability, controlled drug release, and biocompatibility. Polymers used in DDS can be classified into conventional and novel types depending on their qualities and characteristics. While conventional ones, are alginates, cellulose, chitosan, and so forth, have been widely utilized and developed, they have certain disadvantages, such as high temperatures, specialized equipment, complicated processing procedures may be required, high manufacturing costs and energy usage [17]. On the other hand, novel polymers may provide more inventive and efficient processing processes, which can reduce energy consumption, enhance manufacturing efficiency, and result in cost savings. Furthermore, to target specific tissues or cells, new polymers can be coupled with ligands that targets, such as antibodies or peptides. Polymers can also increase drug stability, solubility, bioavailability and efficiency of hydrophobic drugs by solubilizing them. Furthermore, polymers can shield drugs from enzyme breakdown or extreme conditions in the body, extending their shelf life and improving their stability. Polymers have minimal toxicity, high biocompatibility and are simple to produce, making them a desirable choice for DDSs. They offer a lot of potential in fields including treating cancer and genes, immunization, delivery, and adaptability. In general, the implementation of polymer in DDSs has revolutionized the sector and has the potential to increase therapeutic effectiveness and safety, giving patients improved treatment alternatives [18]. Hence, the polymers and recent advancements, such as grafted copolymers, are investigated extensively.

2.2 GRAFTED COPOLYMERS

Grafting has previously been accepted as a legitimate approach for changing the molecular composition of polymers by interacting with synthetic monomers, resulting in a unique biomaterial that is far more sophisticated than its parent polysaccharide. A grafted copolymer is merely the outcome of a polymer and a synthetic monomer reacting to generate a large molecule with a main chain. The monomer is continuously attached as the side chain, making the copolymer. These are intricate macromolecules composed of a main polymer, that is, the backbone of the polymer, connected with side chains of varying polymer composition. The backbone polymer can be simple, branched and or linked together, while the side chains may have different lengths and chemical compositions. Graft polymerization of synthesized polymers on a polysaccharide spine is one of the most effective methods for using polysaccharides in an extended-release DDS. Copolymerization is a simple process that changes the framework of naturally occurring polymers, resulting in desirable biological materials in controlled and sustained release applications. In contrast, native polymers may be unsuitable in DDSs due to their significant swelling and rapid enzymatic degradation in biological fluids [19]. The characteristics of grafted copolymers may be adjusted by varying the constitution and lengths of the side chain and by altering the distribution and density of the grafting points along the backbone polymer. Grafted copolymers are used as DDSs, surface substitutions, compatibilizers, and dispersants, amongst other applications. It is required to adjust the characteristics of polymers to meet the requirements of multiple applications. Blending, Grafting, Curing are all methods for modifying polymer characteristics. Blending is the physical combination of more than one polymer to get the desired characteristics. A bonding of covalent bonds of monomers on the polymer chain accomplishes grafting. Curing entails polymerizing an oligomer mixture to produce a coating that clings to the substratum by physical force, accompanied by finishing by filling in the gaps in the surface [20].

2.2.1 PROPERTIES OF GRAFTED COPOLYMERS

Grafted copolymers have unique properties that are different from those of the individual polymers. Some properties of grafted copolymers are discussed herewith;

2.2.1.1 CHEMICAL PROPERTIES OF GRAFTED COPOLYMERS

Grafted copolymers are more than just physical combinations of two homopolymers; they also have novel chemical characteristics. Only the final stage of the branching chain contributes to the development of grafting area in the case of a branched chain. It may be concluded that grafting has minimal impact on the chemical characteristics of branched-chain polymers; however, numerous places on the main chain remain available for involvement in developing grafting sites. The graft sites occupy multiple locations for reaction on the primary chain, causing considerable variations in properties, mainly chemicals on the main chain and as a result, grafting copolymerization primarily affects the quality of the main chains. It was previously discovered that methacrylic acid (MA) grafting woolen fibers could improve coloring rates and prevent mildew degradation. It was also discovered that when the grafting reaches 10%, the alkaline dissolution of wool is significantly enhanced. Acrylonitrile with cellulose grafted can also considerably increase its microbial resistance and cause alterations in the chemical characteristics of the primary chains [21].

2.2.1.2 SOLUTION PROPERTIES OF GRAFTED COPOLYMERS

Many aspects can influence solution properties; the chemical structure of the graft and backbone polymers, grafting degree, and the molecular weight and concentration of the copolymer in solution. The entire macromolecules are in the ailment extending to the main chain when the grafted copolymers appear to be liquid in a suitable medium for the primary and secondary chains. When one chain is unfavorable solvent, and another is in an appropriate solvent, the chain in the unfavorable solvent shrinks and forms a multimolecular or single micelle disseminated in the medium. The grafted copolymer morphology in various solvents shows that the characteristics of the grafted copolymer were solvent dependent. When the solvent employed is unfavorable for the X section but excellent for the Y section, the X section is scattered and the Y section is constant on the other hand, the Y section is scattered phase, and the X section is constant [21].

2.2.1.3 MECHANICAL PROPERTIES OF GRAFTED COPOLYMERS

A polymer solution forms when the naturally occurring rubber G poly(methyl methacrylate) (PMMA) grafting copolymer dilutes in benzene like common solvent. If methanol is introduced as a solvent to the rubber chain (RC) at this point, the RC contracts, forming the rubber particles covered by the PMMA chain. If methanol replaces with petroleum, a PMMA chain shrinks and produces rubber-coated particles. These two types of colloidal particles are type A and type B. When the solvent evaporates, the rubber particles disperse, and the PMMA is constant in type A, while the PMMA distributes in a continuous phase in type B [22].

2.2.2 TYPES OF POLYMERS USED FOR INNOVATIVE DRUG DELIVERY SYSTEM

Grafted copolymers have many potential applications in pharmacy due to their unique properties and versatility. Depending upon the usage of the polymer it can be synthesized and grafted. Here are some examples of grafted copolymers that includes:

2.2.2.1 NATURAL POLYMERS (CONVENTIONAL POLYMERS)

Carboxymethyl cellulose (CMC) or cellulose gum, also known cellulose is a basic natural form, but it has undergone modification with responsive carboxymethyl clusters linked to the hydroxy groups of its cellulose glucopyranyl moiety to produce a copolymer [23]. Starch is a large glucose atom with glycosidic chains that connect its numerous glucose units. Depending on the source, starch is often composed of varying load rates of amylase (straight and helical) and amylopectin (expanded) particles. Numerous significant alterations were made throughout the period spent on starch production in plants: First, ATP is used to convert glucose-phosphate into adenosine diphosphate-glucose via adenylyl transferase-catalyzed oxidation [24] Pectin is an intriguing heteropolysaccharide structure, and its grafted copolymers offer prospective uses in pharmacy due to their distinct characteristics and adaptability. Pectin can be grafted with chitosan copolymers, which are made up of chitosan, and pectin adjacent chains. Because of their capacity to increase dissolution, drug release kinetics, and biocompatibility they are employed in DDS such as nanoparticles and hydrogels. Other examples are pectin-grafted poly (acrylic acid) (PAA) copolymers that consist of a PAA backbone with pectin side chains attached. They are utilized in pH-sensitive DDSs because the pectin side chains swell and release the drug in reaction to pH changes, and pectin-grafted alginate copolymers, which are made up of alginate, a natural polymer, and pectin side chains linked to it [25]. Chitosan is a type of cationic polysaccharide made from chitin. Chitosan-graft-poly (ethylene glycol) copolymers that have chitosan with PEG side chains and are used in DDSs such as nanoparticles and hydrogels to improve solubility, biocompatibility, and drug release kinetics. Chitosan-graft-poly (acrylic acid) copolymers combine chitosan with PAA side chains and are employed in pH-sensitive DDSs because the PAA side chains swell and release the drug in reaction to pH variations. Chitosan-graft-poly (lactide-co-glycolide) copolymers consists of chitosan with poly (lactide-co-glycolide) (PLGA) side chains attached to it and are used in DDS, such as nanofibers and microspheres, due to their ability to improve drug loading, release kinetics, and stability [26]. A Carrageenan is a collection of gel-like and usually direct polysaccharides that contain sulfated galactose and a 3,6-anhydro D-galactose spine. It may be prepared accordingly to the degree of straight chain sulfation and the level of substitution occurring at the free hydroxy clusters on the direct polysaccharides chain. The subatomic symmetry is malleable, framing weak helical modifications, resulting in most carrageenan existing as gels at ambient temperature. Carrageenan-grafted poly (vinyl alcohol) (PVA) copolymers are made up of carrageenan with PVA chain reactions attached to it and are utilized in systems for DDSs such as microspheres, hydrogels and nanoparticles because they enhance drug loading, release kinetics, and stability [27]. Dextrin is an extremely low sub-atomic weight starch polymer composed primarily of glucose (D) units linked by glycosidic bonds. Because of its capacity to undergo a reversible phase transition in response to temperature variations, dextrin-grafted poly (N-isopropyl acrylamide) (dextrin-g-PNIPAM), copolymer offers potential applications in DDSs such as hydrogels and nanoparticles, and it can deliver various drugs, notably insulin and anticancer drugs. Dextrin-grafted poly (ethylene glycol) (dextrin-g-PEG) copolymer has potential applications in DDSs, such as hydrogels and nanoparticles, due to its ability to improve solubility, biocompatibility, and drug release kinetics. Dextrin-g-PEG copolymer has been used to deliver various drugs, including anticancer agents and proteins [28]. Alginate is a corrosive anionic polysaccharide, and by brushing inherent properties of polysaccharides and proteins or arranging appropriate connections inside macromolecules, alginate can develop innovative capabilities. The combined organization of biopolymers can develop protein platforms with improved health and utility characteristics; because of its capacity to increase drug loading, release kinetics, and stability, alginate-grafted poly (methyl methacrylate) (PMMA) copolymers are employed in DDSs such as hydrogels and nanoparticles. A range of drugs, notably antibiotics, and anti-inflammatory compounds, have been delivered using alginate-PMMA copolymers. Gelatin, a source of inexhaustible resources and a source of calcium, is used to fabricate hydrogel microparticles and forms constrained off-chip by twisting or on-chip by the dissolved dispersal of pregelled beads at a liquid interface [29].

2.2.2.2 SYNTHETIC COPOLYMERS (NOVEL POLYMERS)

Polyethylene glycol (PEG) grafted copolymers consists of a backbone of either polylactide (PLA) or polyglycolide (PGA) with PEG side chains attached. They are used in DDS, such as micelles and nanoparticles, due to their ability to improve solubility, biocompatibility, and drug release kinetics [30]. PAA grafted copolymers consists of a backbone of polyvinyl alcohol (PVA) with PAA side chains attached. They are used as pH-sensitive DDSs, as the PAA side chains can swell and release the drug in response to changes in pH [31]. Poly (N-isopropylacrylamide) grafted copolymers consists of a backbone of either PEG or PVA with poly (N-isopropylacrylamide) (PNIPAM) side chains attached. They are used as thermoresponsive DDSs, as the PNIPAM side chains can collapse or expand in response to changes in temperature, leading to controlled drug release. These copolymers are known for their temperature-sensitive behavior, as they undergo a phase transition from a hydrophilic to a hydrophobic state above a specific temperature (known as the lower critical solution temperature or LCST). This property has led to their application in DDS, where they can be used to deliver drugs at a particular location or in response to a change in temperature [32]. Poly (lactic acid) (PLA) grafted copolymers are formed by grafting lactic acid onto a polymer backbone. These copolymers have potential applications in DDS due to their biodegradability and biocompatibility. PLA-grafted copolymers have been used to deliver various drugs, including anticancer agents and peptides [33]. Poly (ε-caprolactone) (PCL) grafted copolymers are formed by grafting ε-caprolactone onto a polymer backbone. These copolymers have potential applications in DDSs due to their biodegradability and biocompatibility. They can be used to deliver various drugs, including proteins and peptides [34]. Poly (ethylene oxide) (PEO)-grafted copolymers are formed by grafting PEO onto a polymer backbone. These copolymers have been used in DDSs due to their biocompatibility and ability to form micelles, which can encapsulate hydrophobic drugs for targeted delivery [35]. Poly (propylene oxide) (PPO) grafted copolymers are formed by grafting PPO onto a polymer backbone. These copolymers have been used in DDS due to the biocompatibility, hydrophilicity, and ability to form micelles, which can encapsulate hydrophobic drugs for targeted delivery [36].

2.2.3 TECHNIQUE FOR PREPARATION OF GRAFTED COPOLYMERS

2.2.3.1 APPROACHES FOR GRAFTING OF COPOLYMERS

The development of the grafted copolymer can be attributed to three fundamental approaches as shown in Figure 2.2 that is, grafting on, grafting from, and grafting through.

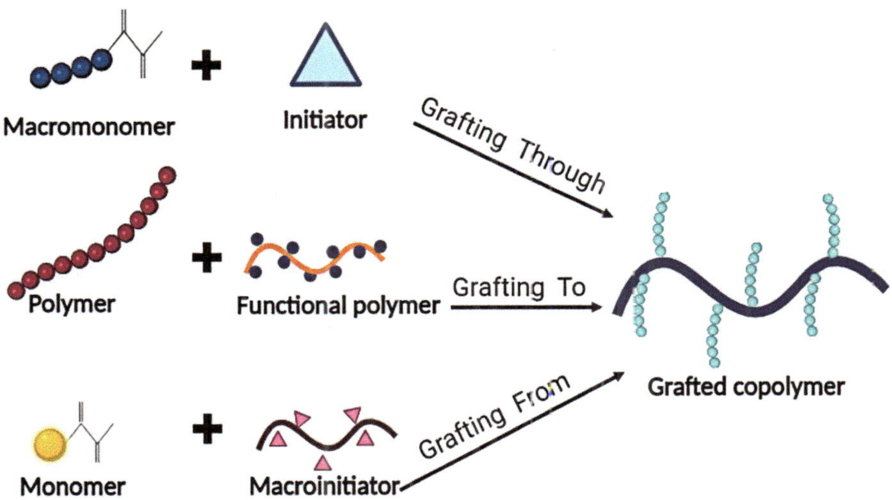

FIGURE 2.2 Common approaches for grafting copolymers: grafting to, grafting from, and grafting through.

The grafting-onto approach is a coupling reaction between polymer-polymer, which involves connecting side chain polymers to a backbone polymer. The chloro (bromo) methylation of polystyrene and subsequent interaction with polymeric chains is a typical technique, and Polystyrene-g-poly (ethylene oxide) grafted copolymers (PS-g-PEO) were synthesized using this approach. Polystyrene (PS) is chloromethylated using a CCl_4 solution containing PS and chloromethyl methyl ether with $SnCl_4$ as a catalyst, and the two chains are pre-synthesized individually; the lengths of the side chains and backbone can be easily modified. A highly effective coupling reaction is needed to prepare graft polymers with large chain densities or longer chain lengths, and the unreacted side chains eliminate due to the poor reactivity of the coupling reaction between polymer-polymer; this can be rectified using effective coupling processes during the synthesis of grafted copolymers [37].

The grafting-from method is based on controlled/living and the grafted copolymer evolved by combining multiple starting units on a backbone polymer. The backbone is chemically changed in the grafting "from" approach to add active sites capable of beginning the polymerization process of a second monomer. The total number of graft chains is determined by the total no. of functioning sites formed along the backbone, presuming that each one contributes to the development of one branch. However, there may be a discrepancy in the lengths of the generated grafts due to kinetics and steric hindrance effects. Poly (methyl methacrylate)-g-(-butyrolactone) copolymers were synthesized using the grafting "from" approach. Anionically polymerized PMMA is subjected to treatment with a potassium hydroxide 18-crown-6 complex, resulting in a macro-molecular activator [37].

The grafting-through approach involves the copolymerization or homopolymerization of macromonomers with a polymerizable group at the end. Various macromonomers are formed by adding polymerizable functional groups to the end of polymer, and by synthesizing macromonomers, the side chains of grafted copolymers may be readily designed. Another benefit is the capacity to add side chains with complete grafting density that none of the other two techniques can achieve. However, the low polymerizable macromonomer is especially deleterious since the reaction occurs between the effectively hindered active center at the end of poly (macromonomer) and the polymerizable group of the macromonomer and, as a result, numerous deactivated controlled/living polymerizations (CRPs) occur, Such as polymerization of live radicals, cannot be utilized to polymerize high molecular weight macromonomers, because of its tolerance for bulky polymerizable groups, ring-opening metathesis. Polymerization is projected to be efficient for the polymerizations of macromonomers as one of the best ways to overcome this steric hindrance [38].

2.2.3.2 TECHNIQUE FOR PREPARATION FOR GRAFTING OF COPOLYMERS

Physical activation entails applying stress to the polymeric backbone and swelling it with appropriate solvents. When stress is applied to the polymeric backbone, it produces segmental movements and molecular flow, which can result in bond scission and generation of free radicals. Active sites are also formed during the polymer monomers freezing and thawing together, whereas milling and mastication of the backbone polymer may create free radicals. In a batch-type internal mixer, dimethylamino-ethylmethacrylate is grafted on linear low-density polyethylene to produce a polymer with primary pendant moieties [39].

Two distinct processes are involved in the chemical approach of activation:

1. **Mechanism of Free Radicals: The Chemical Approach of Activation Employs Two Distinct Mechanisms:** The following processes are involved in the free radical-mediated mechanism for grafting copolymers: The production of radicals free from activator molecules, such as azo compound or peroxide, which initiate the reaction and stimulation can be achieved by thermal or photochemical reactions. The free radicals attacking monomer units of grafted copolymer produce an additional free radical on the monomer unit. This procedure is repeated, resulting in graft chain growth. Ultimately, combining two free radicals or incorporating an atom of hydrogen into an unstable radical terminates the process [40].
2. **Polymerization Via Ionization Process:** Grafted copolymers can also be made via ionic polymerization, for example, ionic transfer of chains between polystyrene level-based carbon cations and in poly (p-methoxy styrene-g-Polystyrene). Furthermore, aluminum bromide can stimulate the carbocation polymerization of isobutylene to generate the Polystyrene polyisobutylene grafted copolymers in carbon disulfide containing chloromethylated polystyrene and, if chlorine polymers (e.g., PVC) combine with aluminum trichloride or ferric tetrachloride macromolecular carbon ions are produced, which can be used to start the copolymerization of styrene monomers. Anionic polymerization can also be used to make grafted polymers, for example, the aryl groups found in ethylene polymer with aromatic ring structures can react with metals like alkali resulting in a colored charged transferring complex; it can start the process of polymerization of epoxides and forms grafted copolymers [40].

The radiation method that utilizes UV radiation may stimulate polymers and produce sites of activity for grafting as the absorbed energy is low when exposed to ultraviolet rays; it is appropriate for surface grafting. The photosensitive initiators or activation of dyes that are sensitive to UV can considerably boost the yield of grafted polymers. UV radiation can also introduce H_2O_2 groups into skeletal polymers in the presence of air, and the skeleton polymer chain can also contain photosensitive groups. High-energy radiation technologies are commonly employed in practical applications; for example, gamma ray radiation utilizing Co60 as a radiation source has been extensively researched and is widely engaged in industrial production. There are several techniques for producing grafted polymers the most common is using gamma radiation, and the most basic way is to apply direct radiation to the polymer-monomer combination in a gas or liquid phase. When the polymers are unsuitable with the monomers or can't be swollen by the monomers, the process of grafting frequently occurs only on the surface of the polymer [41].

2.4 FACTORS AFFECTING GRAFTING OF POLYMERS

The grafting of copolymers involves attaching one polymer chain to another polymer backbone. The process is influenced by several factors, including:

2.4.1 EFFECT OF MONOMER

The amount of time spent integrating the monomer is also critical and depends on variables such as steric and polarity, spine swellability in close proximity to monomers, and monomer convergence. Grafting productivity increased with the expansion of the monomer targeted up to a certain maximum and then decreased with additional progressive monomer fixation. This pattern of behavior may reflect a considerable increase in monomer centralization along the spine. After the cut-off, an increase in monomer convergence speeds up the homo-polymerization reaction rather than the unification. The chemical structure of grafting monomer can influence efficiency of grafted copolymers. For instance, greater reactivity monomers, including acrylates and methacrylates, can produce more efficient grafted copolymers because they tend to polymerize and graft onto the substrate easily. Monomers with lesser reactivity, such as styrenes, may result in poorer effectiveness due to delayed or fewer efficient grafting processes [42].

2.4.2 EFFECT OF INITIATOR

All synthetic uniting reactions demand an initiator and its fixation, nature, work, and cosolvency. Azobisisobutyronitrile (AIBN), and Potassium persulfate ($K_2S_2O_8$) are examples of such initiators, where reverberation adjustment and on other hand ordinary peroxide initiators do not have as much reverberation adjustment as AIBN and, therefore, peroxide initiators have a greater uniting yield [43].

2.4.3 EFFECT OF SOLVENT

In the grafting mechanism, a solvent is a medium that transports monomers to the surrounding region of the backbone. Numerous characteristics influence the selection of an appropriate solvent, including the solubility of the monomer in the solvent, the swelling properties of the backbone, the compatibility of the solvents being used (if more than one is employed), and the formation of free radicals in the presence of the solvent. The nature of solvent significantly impacts the ability to dissolve monomers and polymers; for example, alcohols can extend the structural framework and break styrene, enabling its monomer to permeate throughout the cellulose structure readily. The degree of grafting progressively reduces each alcohol bonding, changing methanol to isopropanol and finally to t-butanol due to the continuously decreasing swelling capacities of alcohol [44].

2.4.4 EFFECT OF TEMPERATURE

The energy of unified copolymerization is regulated by temperature. The yield rises when the temperature rises until a new point is reached. Monomeric dissemination, which gradually increases with increasing temperature in the spine, is one reason for favoring faster grafting. When the heat rate rises of the initiators and free radicals are formed, the effectiveness of the initiator on the base polymer increases with increasing temperature, resulting in increased polymer macroradical fixation and, thus, improved polymerization. The polymerization kinetics of grafted copolymers can also be affected by temperature; for example, higher temperatures can result in quicker polymerization rates, resulting in increased graft density and short graft chains, and the temperature can also affect the polymerization kinetics during the synthesis of grafted copolymers, for example, higher temperatures can generally lead to faster polymerization rates, for example, PNIPAM, resulting in higher grafting densities and shorter graft chains, conversely, lower temperatures can lead to slower polymerization rates, such as PEG, resulting in reduced grafting density and longer graft chains, it can impact the overall structure and morphology of grafted copolymers [45]. Factors affecting the efficiency of grafted copolymers are shown in Figure 2.3.

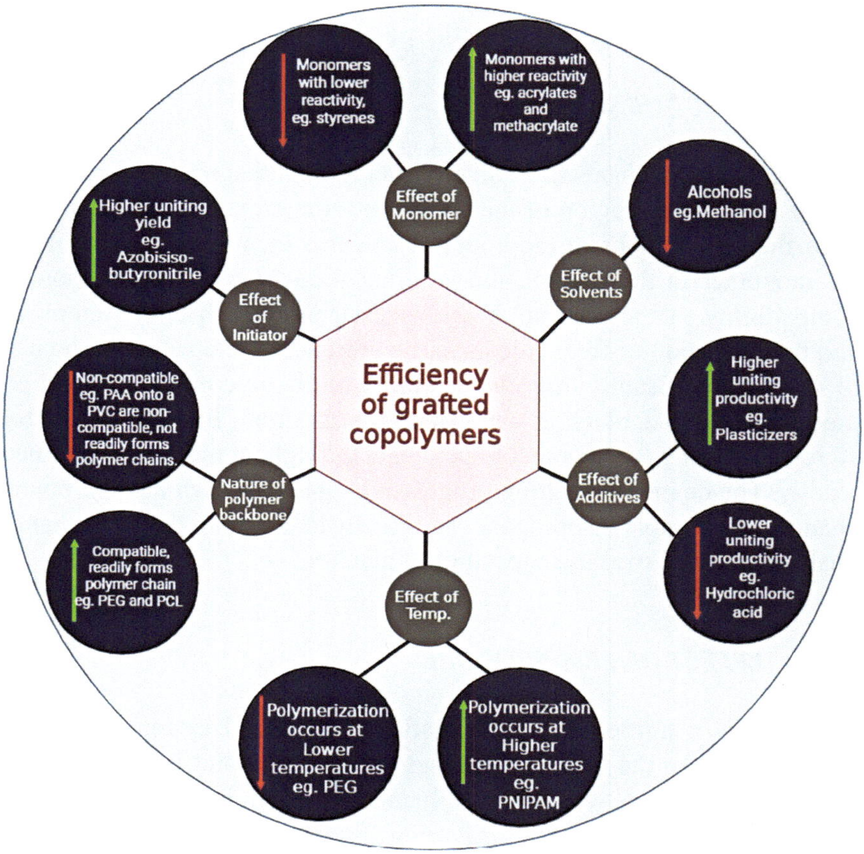

FIGURE 2.3 Factors affecting the efficiency of grafted copolymers.

2.4.5 EFFECT OF ADDITIVES

The uniting yield or level of joined copolymerization is controlled by additives such as metal particles, acids and inorganic salts. The reaction between the monomer and the backbone must compete with any responses between the monomer and the additional substances. Although a few different compounds may increase the spine/monomer reaction by increasing grafting productivity, a turnaround will be genuine if the response between monomer and added substance is prominent; for example, chlorine is consolidated in the polymer, and chlorine has been observed to have a detrimental effect on uniting productivity [46]. Additives can also be used to improve the performance of polymers in specific applications; for example, plasticizers can improve the elasticity and hardness of polymer compounds, making them more suited for applications requiring flexibility, such as flexible packaging films or medical equipment [46].

2.4.6 NATURE OF POLYMER BACKBONE

The physical and chemical nature of the polymer backbone plays a vital role in grafting, including ionic linkage among monomers and the formed polymer backbone. Suppose a hydrophilic polymer such as cellulose is added to water to initiate grafting; it fails to initiate grafting processes due to its insolubility and due to the large size of the polymeric chain, linking amino acids, cystine bonds, and intramolecular H-bonding. When the same polymer is exposed to UV light, oxidative processes begin, and free radicals develop, eventually leading to grafting when monomers are present [47].

2.5 CHARACTERIZATION OF GRAFTED COPOLYMERS

Characterization of grafted copolymers involves a comprehensive analysis of their physical, chemical, and structural properties. This characterization is typically performed using various techniques to gain a detailed understanding of the characteristics of the copolymer. For confirmation of polymer grafting, the following test can be employed.

2.5.1 MOLECULAR WEIGHT

Molecular weight characterization employs gel permeation chromatography (GPC), a widely used technique for characterizing grafted copolymers. GPC separates polymers according to their size and molecular weight, which is important for understanding the structure and properties of grafted copolymers. In GPC, a sample of the grafted copolymer is dissolved in a suitable solvent and injected into a column packed with a gel matrix. The gel matrix contains pores of a specific size, which separates the polymers based on their size, and as the sample travels through the column, smaller molecules penetrate deeper into the pores and are eluted later than larger molecules. The elution time of each polymer fraction is measured using a detector, typically a refractive index detector, and a chromatogram is generated. GPC is a powerful technique for characterizing grafted copolymers and can provide valuable information for developing new materials and applications [48].

2.5.2 CHEMICAL IDENTIFICATION AND MATERIAL CHARACTERIZATION

Fourier transform infrared spectroscopy (FTIR) may be employed to identify particular functional groups in compounds, and all grafted copolymers can be produced in 0.5–1 mm thick film for characterization, with that attenuated total reflectance can be used to assess the results using the transmittance mode for FTIR. For grafted copolymers, FTIR can be used to determine the presence of the grafting monomers, the degree of grafting, and the chemical structure of the grafted copolymer backbone. FTIR can also identify any functional groups introduced during the grafting process, such as reactive groups used for crosslinking or coupling agents [49].

2.5.3 GRAFTING PERCENTAGE

It is the amount of weight increase of native cellulose after graft with a monomer and is determined [50] using the following equation:

$$\frac{W1 - W0}{W0} \times (100)$$

Where W0 and W1 are the cellulose original cellulose and grafted copolymer weights, respectively.

2.5.4 GRAFTING EFFICIENCY

The percentage of monomer grafted onto cellulose divided by the quantity of monomer transformed to grafted polymers with the homo-polymers, or the proportion of polymer grafted to cellulose as a whole polymer, and it is computed using the following equation [51]:

$$\frac{W1 - W0}{W1 - W0 + W2} \times (100)$$

where W2, W1, W0, and are the homopolymer weights, original cellulose, cellulose grafted copolymer, and respectively. Subtracting the quantity of grafted polymer plus the quantity of unreacted monomer from the original amount of monomer yields the weight of homopolymer (W2).

2.5.5 WATER ABSORPTION CAPACITY

It refers to the ability of a copolymer to absorb and retain water when exposed to a liquid or humid environment when ionic or hydrophilic polymers such as 2-acrylamido-methyl propane sulfonic acid, acrylamide or their binary mixtures are grafted onto cellulose, and the resultant copolymer absorbs a large amount of water depending on the grafting percentage, the ionization graft chain length, as well as swelling medium and ionic strength, for example, the capacity of PAA-grafted cellulose microfibers absorb more water and, have shown 3x times more capacity than original cellulose microfibers [52]. The following equation can be used to calculate the ability to absorb water of cellulose-grafted copolymers:

$$\text{Water absorption capacity} = \frac{Weight\ of\ swollen\ copolymer\ (Ws) - Weight\ of\ dry\ copolymer\ (Wd)}{Weight\ of\ dry\ copolymer\ (Wd)}$$

2.5.6 STRUCTURE CHARACTERIZATION

Nuclear magnetic resonance (NMR) spectroscopy is a powerful analytical technique that can characterize the structure, composition, and properties of grafted copolymers. NMR spectroscopy provides valuable information about the chemical environment and connectivity of different atoms in a molecule, which can help elucidate the structure and properties of the grafted copolymer [53].

2.5.7 CRYSTALLOGRAPHIC PROPERTIES

X-ray diffraction (XRD) is a powerful technique used to study crystallography, and it can provide data mainly on the crystal structure, crystallinity, and morphology of the grafted copolymers [54].

2.5.8 THERMAL PROPERTIES

Differential Scanning Calorimetry (DSC) is a commonly used thermal analysis technique that can provide helpful insights of thermal properties and behavior, and it also measures the heat flow into or out of a sample as a function of temperature or time, allowing for the determination of various thermal parameters, such as glass transition temperature (Tg), melting temperature (Tm), crystallization temperature, and thermal stability, which are essential for understanding the properties and behavior of grafted copolymers [55].

2.5.9 VISCOSITY MEASUREMENTS

Rotational viscometry, Capillary viscometry and Oscillatory viscometry are some common techniques to characterize flow behavior and rheological properties of grafted copolymers. Viscosity measures resistance of flow and can provide insights and data about the molecular structure, composition, and processing behavior of grafted copolymers [56].

2.5.10 SWELLING STUDIES

It is commonly performed to investigate the swelling behavior and properties of grafted copolymers, which are important for understanding their physicochemical characteristics and potential applications. Swelling studies can provide insights into the responsiveness of grafted copolymers to environmental changes, such as changes in solvent, temperature, pH, or other external stimuli. Some common methods used for swelling studies of grafted copolymers include gravimetric swelling studies, optical swelling studies, and dynamic mechanical analysis [57].

2.5.11 BIODEGRADABILITY

The biodegradability of the copolymer relies on the monomers bonded onto cellulose. Isocyanate-terminated poly (caprolactone-2-monoacrylate) (NCO-PCLA) on the cellulose diacetate (CDA) shows that grafted copolymers have excellent biodegradability in natural conditions using soil burial tests as well as active sludge tests [58]. The length and structure of the grafts can also affect the biodegradability of grafted copolymers, for example, longer grafts may hinder the accessibility of enzymes or microorganisms responsible for biodegradation and thus reduce the biodegradation rate; similarly, the structure of the grafted copolymers, such as their chemical composition and configuration, can impact their biodegradability. Biodegradability testing and evaluation of grafted copolymers typically involves conducting standardized tests, such as soil burial tests, to determine their biodegradability under specific conditions. It is recommended to consult scientific literature or conduct specific biodegradation testing for a particular grafted copolymer to assess its biodegradability accurately [58].

2.6 CURRENTS TRENDS

Grafted copolymer side chains can have different chemical compositions, which allows for a wide range of applications. Here are some innovations of grafted copolymers.

2.6.1 GRAFTED COPOLYMERS FOR NANOPARTICULATE DRUG DELIVERY SYSTEM

Grafted copolymers have gained significant attention in the field of nanoparticulate DDSs due to their unique and advantageous properties.

Grafted copolymers provide enhanced drug encapsulation and controlled release capabilities. The grafting of functional polymers onto a base polymer can develop unique microstructure, such as core-shell or micellar structures, that gives improved drug loading efficiency and stability. The grafting of responsive or stimuli-sensitive polymers can also enable triggered drug release in response to specific environmental cues, such as pH, temperature, or enzymatic activity, which can enhance the therapeutic efficacy of the drug [57].

Grafted copolymers can offer improved biocompatibility and safety for drug delivery applications. The selection of appropriate grafting polymers can improve the biocompatibility of the copolymer, reducing potential toxicity and immunogenicity. Grafted copolymers can also provide a protective shield around the drug-loaded nanoparticles, reducing premature drug release and minimizing off-target effects, which can enhance the safety and efficacy of the DDS [56].

Grafted copolymers can enable targeted drug delivery to specific cells, tissues, or organs. The grafting of ligands, peptides, or targeting moieties onto the copolymer can impart specificity and selectivity to the DDS, allowing for targeted delivery of the drug to the desired site of action. This can enhance the therapeutic efficacy of the drug while minimizing systemic side effects [57].

Grafted copolymers can provide improved stability and protection to the encapsulated drug. The grafting of protective polymers or surfactants can shield the drug from degradation, oxidation, or other adverse conditions, preserving its stability and bioactivity. This can extend the shelf-life of the drug-loaded nanoparticles and improve their performance during storage, transportation, and administration [58].

2.6.2 GRAFTED COPOLYMERS FOR VESICLE MEDIATED DRUG DELIVERY SYSTEM

Here are some specific aspects of novelty for grafted copolymers in vesicle DDSs:

Grafted copolymers can improve the stability and drug-loading capacity of vesicles. The grafting of functional polymers onto the vesicle surface can enhance the stability of the vesicles by preventing their aggregation, fusion, or leakage, thereby improving the shelf-life and performance of the DDS. Grafted copolymers can also increase the drug-loading capacity of vesicles by providing additional space for drug encapsulation, reducing drug leakage, or facilitating drug encapsulation through specific interactions [59].

Grafted copolymers can enable controlled drug release from vesicles. Grafting responsive or stimuli-sensitive polymers onto the vesicle surface can enable triggered drug release in response to environmental cues, such as pH, temperature, or enzymatic activity, leading to controlled and sustained drug release. This can improve the therapeutic efficacy of the drug and reduce the frequency of dosing [59].

Grafted copolymers can improve the biocompatibility and targeting capabilities of vesicles. The grafting of biocompatible polymers onto the vesicle surface can reduce potential toxicity and immunogenicity, making the vesicles more suitable for *in vivo* drug delivery. Moreover, grafting targeting ligands, peptides, or antibodies onto the vesicle surface can impart specificity and selectivity to the DDS, allowing for targeted delivery of the drug to specific cells, tissues, or organs [60].

Grafted copolymers allow for the tailoring of vesicle properties to meet specific requirements. The selection of suitable grafting polymers can modify the vesicles' size, surface charge, and membrane permeability, which can affect their drug encapsulation, stability, and biodistribution properties. This flexibility in vesicle design enables the development of vesicle DDSs with optimized properties for specific drugs or therapeutic applications.

Grafted copolymers offer the opportunity to combine different polymer properties in a single vesicle system. The grafting of multiple types of polymers onto the vesicle surface can provide synergistic effects, such as improved stability, enhanced drug loading capacity, and controlled drug release, leading to unique and superior drug delivery performances that are not achievable with single-component vesicles [60].

2.6.3 GRAFTED COPOLYMERS FOR MICROELECTROMECHANICAL APPLICATIONS

Grafted copolymers have shown novelty and potential in Microelectromechanical Systems (MEMS), which are miniaturized devices that combine electrical and mechanical components on a single chip. Here are some specific aspects of novelty for grafted copolymers in MEMS:

Grafted copolymers can modify the surface properties of MEMS devices to achieve specific functionalities. The grafting of functional polymers onto MEMS surfaces can provide desirable surface properties, such as improved hydrophobicity, hydrophilicity, biocompatibility, or anti-fouling properties. These modified surface properties can enhance the performance and reliability of MEMS devices by reducing stiction, improving sensor response, or preventing biofouling [61].

Grafted copolymers can act as a passivation layer on the surface of MEMS devices. The grafting of passivating polymers onto MEMS surfaces can protect them from environmental factors such as moisture, chemicals, or contaminants, which can degrade the performance and longevity of MEMS devices. The passivating effect of grafted copolymers can help prevent corrosion, wear, or degradation of MEMS devices, leading to improved device reliability and longevity [62].

Grafted copolymers can tune the mechanical properties of MEMS devices. The grafting of polymers with specific mechanical properties, such as flexibility, elasticity, or damping, onto MEMS structures can alter their mechanical behavior, leading to improved device performance. For example, grafted copolymers can be used to reduce stress concentrations, enhance mechanical stability, or provide damping for MEMS resonators, sensors, or actuators.

Grafted copolymers can enable selective surface functionalization of MEMS devices. The grafting of functional polymers with specific chemical or biological functionalities onto MEMS surfaces can enable selective binding or immobilization of target molecules, such as sensing agents, enzymes, or antibodies. This selective surface functionalization can enable specific sensing, detection, or bioassay capabilities in MEMS devices, making them more versatile and applicable in various applications, such as biosensors, microfluidic devices, or lab-on-a-chip systems [63].

Grafted copolymers can integrate multiple functionalities into a single MEMS device. Grafting different types of polymers with complementary properties, such as electrical, mechanical, or chemical functionalities, onto different parts of a MEMS device can enable multifunctionality in a single device. This can lead to compact, miniaturized MEMS devices with enhanced performance and versatility, as multiple functions can be integrated into a single chip [64].

2.6.4 GRAFTED COPOLYMERS FOR OPTOELECTRONIC APPLICATIONS

Grafted copolymers have shown significant potential in optoelectronic applications in the field of pharmacy. Here are some aspects of their novelty and potential in this specific context:

Grafting of polyacrylamide onto poly (aniline) (PANI) by electrospinning technique produces fibers matting for supercapacitor applications, which is an intriguing method for grafting copolymers [65]. On the other hand, grafted copolymers consisting of different attached cores, where poly (3-hexylthiophene) (P3HT) arms are transplanted onto the thiophene (Th) moieties of the cores, the Majority of these donor-acceptor copolymers exhibit poor performance when applied in organic field-effect transistors (OFET) topologies. It displays low on/off ratios of current, indicating inefficient switching between the "on" and "off" states of the transistors. Additionally, these copolymers exhibit very high threshold voltages, implying that a significant voltage is required to activate the transistor [66].

Grafted copolymers can be designed and utilized as DDSs in optoelectronic applications in pharmacy. For example, grafting functional groups onto the main polymer chain can provide drug-loading sites, allowing for the controlled release of drugs in response to specific optical stimuli, such as light irradiation. This can enable precise spatiotemporal control over drug release, which is highly desirable in DDSs for targeted and controlled drug therapy. Grafted copolymers can be designed with photoresponsive moieties that undergo structural changes upon exposure to light, leading to drug release. This can be achieved through various mechanisms, such as photoisomerization, photocleavage, or photo-crosslinking, which specific wavelengths or intensities of light can trigger. This photoresponsive drug release feature of grafted copolymers can enable on-demand drug delivery in response to optical stimuli, providing a highly localized and controlled drug release in pharmacy applications [66].

Grafted copolymers can improve the drug-loading capacity and stability of DDSs. The grafting of functional groups onto the main polymer chain can create additional drug-binding sites, leading to increased drug-loading capacity. Grafted copolymers can also provide enhanced stability to encapsulated drugs by protecting them from degradation or premature release. This can result in more effective and stable DDSs in pharmacy applications. Grafted copolymers can be designed to be biocompatible and biodegradable, which are important characteristics for DDSs in pharmacy applications. Biocompatible grafted copolymers can minimize adverse effects, such as inflammation or toxicity when used *in vivo*. Biodegradable grafted copolymers can degrade over time into biocompatible byproducts, reducing the need for additional removal or extraction procedures. This makes grafted copolymers attractive for developing safe and effective DDSs in pharmacy applications [67].

Grafted copolymers offer versatility in the formulation and administration of DDSs. Depending on the pharmacy application requirements, they can be formulated into various forms, such as nanoparticles, micelles, hydrogels, or films. They can also be administered through different routes, such as oral, topical, transdermal, or injectable, providing flexibility in drug delivery options. This versatility in formulation and administration allows for the customization of DDSs for different types of drugs, patient populations, and administration routes in pharmacy applications [67].

2.6.5 BIOMIMETIC AND BIO-INSPIRED GRAFTED COPOLYMERS FOR MEDICAL APPLICATIONS

Biomimetic and bio-inspired grafted copolymers are a class of copolymers that mimic or are inspired by natural systems in medical applications. Here are some aspects of their novelty and potential in this specific context:

Biomimetic and bio-inspired grafted copolymers can be designed to mimic the chemical and physical properties of natural biomolecules, such as proteins or carbohydrates, which are inherently biocompatible. This can make them well-suited for medical applications where biocompatibility is critical, such as in DDSs, implants, or tissue engineering. By mimicking the natural biomolecules, these copolymers can exhibit enhanced biocompatibility, reducing the risk of adverse reactions and improving patient safety. Biomimetic and bio-inspired grafted copolymers can be designed to mimic specific biological targeting mechanisms, such as receptor-ligand interactions, which can enable targeted drug delivery. For example, the grafted copolymers can be engineered to have ligands on their surface that can specifically recognize and bind to receptors on the target cells or tissues, allowing for site-specific drug delivery. This can enhance the therapeutic efficacy of drugs while minimizing off-target effects, making them ideal for pharmacy applications where precise drug targeting is desired [68].

Biomimetic and bio-inspired grafted copolymers can be designed to respond to specific stimuli, such as pH, temperature, enzymes, or light, in a manner similar to natural systems. This can enable controlled drug release in response to the specific physiological or environmental conditions of the target site. For example, pH-responsive grafted copolymers can release drugs in acidic environments, such as tumor tissues, while remaining stable in normal physiological pH conditions. This can result in controlled and sustained drug release, improving the therapeutic outcomes in pharmacy applications. Biomimetic and bio-inspired grafted copolymers can be designed with surface properties that mimic those of natural biomolecules, such as cell adhesion, signaling, or tissue regeneration. This can enable them to interact with biological systems in a similar manner, facilitating cell attachment, proliferation, and tissue integration. For example, grafted copolymers can be designed to mimic the extracellular matrix, promoting cell adhesion and tissue regeneration in pharmacy applications, such as wound healing or tissue repair [69].

2.6.6 PATENTS

The patent of grafted copolymers is based on process patents describing the methods for synthesizing grafted copolymers. It also includes detailed information on the reaction conditions, the type of monomers used, and the initiators or catalysts employed. Composition patents describe the composition of grafted copolymers and include information on the backbone polymer type, the grafted polymer chains, and the relative amounts of each. Application-based patents describe the use of grafted copolymers in specific applications and include information on the properties of the grafted copolymers that make them suitable for the particular application. Some of the patents are described in Table 2.1.

TABLE 2.1 Recent Patents for Grafted Copolymers

Sr. No.	Year	Patent Number	Title and Description	References
1	2023	US20230087601	**Cosmetic compositions containing low molecular weight amorphous grade polylactic acid resin.** Some polylactic acid polymers are used as film-formers in cosmetic formulations. The lactic acid polymer resin is efficient at low concentrations in the mixture and is suitable for a wide range of organic components that are widely used in cosmetic formulations. Cosmetic formulations can be in the form of a cream, lotion, paste, or spray.	[70]
2	2023	EP4136173	**Fast hardening, aqueous coating compositions containing grafted copolymer of polyamine and vinyl monomers.** An anionically-stabilized aqueous emulsion composition includes (a) a first emulsion polymer having a Tg from -60°C to 40°C, (b) a volatile base, and (c) polymeric particles.	[71]
3	2022	CN114703067	**Photosynthetic microorganism as well as application and plasmid thereof.** The genetic engineering blue algae can be used for efficiently producing polylactic acid (PLA) and lactic acid copolymers by directly utilizing greenhouse gas CO_2 through photosynthesis so that the consumption of carbohydrates and expensive precursors and the use of inducers are avoided.	[72]
4	2022	JP2022130521	**Novel polymers and DNA copolymer coatings.** An acrylamide-based copolymer comprises an azido group-containing acrylamide derivative, and an amine hydrochloride-containing acrylamide derivative provides polymer coatings for surface functionalization and for grafting pre-grafted DNA-copolymers to surfaces of substrates for use in DNA sequencing and other diagnostic applications.	[73]
5	2021	CN113797399	**Application of non-release type antimicrobial adhesion coating in an antibacterial medical instrument.** The invention relates to application of a non-release type antimicrobial adhesion coating in antibacterial medical instruments, the main component of the coating is composed of copolymers of two monomers of bornyl acrylate and polyethylene glycol diacrylate, an amphiphilic three-dimensional chemical antibacterial polymer coating is constructed, and the coating has good antimicrobial adhesion. The preparation method of the coating is simple, and the obtained grafted coating is safe and stable; respective advantages of the bornyl polyacrylate and the polyethylene glycol diacrylate can be fully exerted, and microbial adhesion can be synergistically prevented.	[74]

TABLE 2.1 *(Continued)*

Sr. No.	Year	Patent Number	Title and Description	References
6	2021	TH1901000993A	**Method Of Preparing a Grafted Copolymer of Lignin And/ Or Cellulose.** A method of preparation of grafted copolymers of lignin and/or cellulose, comprising the generation of macroradicals by the mechanical impact on the backbone of said lignin and/or cellulose of said lignin and/or cellulose-containing material and grafting monomers and/or polymers to said backbone.	[75]
7	2020	US20220409654	**Compositions for the treatment of epithelial lesions.** It describes the drugs for the treatment of lesions and wounds of the skin and mucosa. A composition comprising: carbomer (poly acrylic acid) in the range from 0.05% to 5%; sodium hyaluronate in the range from 0.01% to 15%; poloxamer (copolymers of ethylene oxide and propylene oxide) in the range from 0.1% to 15% was developed.	[76]
8	2020	WO/2022/120764	**Hybrid surface coating and preparation method, therefore, and application thereof.** The hybrid surface coating is a coating comprising hard particles and soft particles which is on the surface of a matrix; the soft particles comprise at least one type of temperature-sensitive hydrogel spheres, and the temperature-sensitive hydrogel spheres are selected from poly (N-isopropyl acrylamide) hydrogel spheres, poly (N,N-diethylacrylamide) hydrogel spheres, hydrogel spheres of poly (N-isopropylacrylamide) copolymers, or microspheres prepared by blending poly (N-isopropylacrylamide) copolymer hydrogels and other hydrogels. By means of the coating, cell behavior, and gene expression can be regulated and controlled.	[77]
9	2019	MX2019011721	**Novel Amphiphilic Graft Copolymers.** The addition of hydrophilic PEO side chains will result in amphiphilic properties. The newly developed amphiphilic grafted copolymer is further explained in terms of its applications, that includes, but are not limited to, thermoplastic elastomers, films, fibers, fabrics, gels, ventilated packaging supplies, additives over biodegradable framework, surfactant used, solid polymer electrolytes, antistatic additives, phase transfer catalysts polymers compatibilizer, and biocompatible polymers.	[78]
10	2018	CN107903348	**Preparation method of polyethylene glycol grafted polymethyl methacrylate amphiphilic grafted copolymer.** The preparation method has beneficial effects and provides the method of the polyethylene glycol grafted polymethyl methacrylate amphiphilic grafted copolymer,	[79]

2.7 CONCLUSION

Grafted copolymers have the potential to play a significant role in tissue engineering and regenerative medicine. They can be designed to mimic the extracellular matrix and provide a supportive scaffold for cell attachment, proliferation, and tissue regeneration. Future prospects may involve further customization of grafted copolymers for specific tissue types, such as bone, cartilage, skin, or organs, to promote tissue regeneration and repair. Grafted copolymers may also be combined with other biomaterials or bioactive molecules to create composite scaffolds with enhanced properties for tissue engineering applications. Advancements in surface modification techniques using grafted copolymers to achieve improved properties, such as enhanced biocompatibility, reduced fouling, increased stability, or controlled release of bioactive molecules. Grafted copolymers may also be designed to have specific functionalities, such as antimicrobial properties, cell adhesion, or controlled release of bioactive molecules, for specific applications in healthcare and biomedical fields. Grafted copolymers hold significant promise for a wide range of future applications in areas such as advanced drug delivery, tissue engineering, surface modification, environmental applications, and advanced materials/nanotechnology. Continued research and development in this field are expected to yield further advancements and innovations, leading to new and exciting applications for grafted copolymers in various fields.

KEYWORDS

- **gel permeation chromatography**
- **grafted polymers**
- **methacrylic acid**
- **nanoparticles**
- **poly (acrylic acid)**
- **polyvinyl alcohol**

REFERENCES

1. Tiwari, G., Tiwari, R., Bannerjee, S., Bhati, L., Pandey, S., & Pandey, P. (2012). Drug delivery systems: An updated review. *International Journal of Pharmaceutical Investigation, 2*(1), 2.
2. Faheem, A., & Abdelkader, D. H. (2020). Novel drug delivery systems [Internet]. Sure.sunderland.ac.uk. Elsevier. Retrieved from: https://sure.sunderland.ac.uk/id/eprint/11524/ (accessed on 17 September 2024).
3. Patra, J. K., Das, G., Fraceto, L. F., Campos, E. V. R., Rodriguez-Torres, M. del P., & Acosta-Torres, L. S. (2018). Nano-based drug delivery systems: Recent developments and future prospects. *Journal of Nanobiotechnology, 16*(1). https://jnanobiotechnology.biomedcentral.com/articles/10.1186/s12951-018-0392-8.
4. Su, H., Wang, Y., Liu, S., Wang, Y., Liu, Q., & Liu, G. (2019). Emerging transporter-targeted nanoparticulate drug delivery systems. *Acta Pharmaceutica Sinica B, 9*(1), 49–58.
5. Balachandran, B., & Yuana, Y. (2019). Extracellular vesicles-based drug delivery system for cancer treatment. In U. Schumacher (Ed.), *Cogent Medicine, 6*(1).
6. Kuotsu, K., Karim, K., Mandal, A., Biswas, N., Guha, A., & Chatterjee, S. (2010). Niosome: A future of targeted drug delivery systems. *Journal of Advanced Pharmaceutical Technology & Research, 1*(4), 374–380.
7. Chauhan, N., Vasava, P., Khan, S. L., Siddiqui, F. A., Islam, F., & Chopra, H. (2022). Ethosomes: A novel drug carrier. *Annals of Medicine & Surgery, 82*, 104595.
8. Umashankar, M. S., Sachdeva, R. K., & Gulati, M. (2010). Aquasomes: A promising carrier for peptides and protein delivery. *Nanomedicine: Nanotechnology, Biology and Medicine, 6*(3), 419–426.
9. Garg, G., Saraf, S., & Saraf, S. (2007). Cubosomes: An overview. *Biological & Pharmaceutical Bulletin, 30*(2), 350–353.
10. Nita, L. E., Ghilan, A., Rusu, A. G., Neamtu, I., & Chiriac, A. P. (2020). New trends in bio-based aerogels. *Pharmaceutics, 12*(5), 449.
11. Dias, A. P., da Silva Santos, S., da Silva, J. V., Parise-Filho, R., Igne Ferreira, E., & Seoud, O. E. (2020). Dendrimers in the context of nanomedicine. *International Journal of Pharmaceutics, 573*, 118814.
12. Zhang, L., Gu, F., Chan, J., Wang, A., Langer, R., & Farokhzad, O. (2007). Nanoparticles in medicine: Therapeutic applications and developments. *Clinical Pharmacology & Therapeutics, 83*(5), 761–769.
13. Wei, L., Sun, P., Nie, S., & Pan, W. (2005). Preparation and evaluation of SEDDS and SMEDDS containing carvedilol. *Drug Development and Industrial Pharmacy, 31*(8), 785–794.
14. Forouzandeh, F., Arevalo, A., Alfadhel, A., & Borkholder, D. A. (2021). A review of peristaltic micropumps. *Sensors and Actuators A: Physical, 326*, 112602.
15. Hasan, M., Khatun, A., & Kogure, K. (2022). Iontophoresis of biological macromolecular drugs. *Pharmaceutics, 14*(3), 525.
16. Heilmann, K. (1983). Innovations in drug delivery systems. *Current Medical Research and Opinion, 8*(sup2), 3–9.
17. Ahsanur Reza, M., Quadir, M. A., & Haider, S. (2003). Comparative evaluation of plastic, hydrophobic and hydrophilic polymers as matrices for controlled-release drug delivery. *Journal of Pharmaceutical Sciences, 6*(2), 282–291.

18. Lamprou, D. A., Singh, T. R. R., Larrañeta, E., & Donnelly, R. F. (2019). How innovative drug delivery devices can help realize clinical utility of new effective therapies. *Expert Opinion on Drug Delivery, 16*(12), 1277–1281.
19. Peter, A., Obot, I. B., & Sharma, S. K. (2015). Use of natural gums as green corrosion inhibitors: An overview. *International Journal of Industrial Chemistry, 6*(3), 153–164.
20. Solomon, M. M., Umoren, S. A., Udosoro, I. I., & Udoh, A. P. (2010). Inhibitive and adsorption behavior of carboxymethyl cellulose on mild steel corrosion in sulphuric acid solution. *Corrosion Science, 52*(4), 1317–1325. https://www.academia.edu/18640956/Inhibitive_and_adsorption_behaviour_of_carboxymethyl_cellulose_on_mild_steel_corrosion_in_sulphuric_acid_solution (accessed on 17 September 2024).
21. Gürdağ, G., & Sarmad, S. (2013). Cellulose graft copolymers: Synthesis, properties, and applications. In *Polysaccharide Based Graft Copolymers* (pp. 15–57).
22. Husted, K. E. L., Herzog-Arbeitman, A., Kleinschmidt, D., Zhang, W., Sun, Z., & Fielitz, A. J. (2022). Pendant group modifications provide graft copolymer silicones with exceptionally broad thermomechanical properties. *ACS Central Science, 9*(1), 36–47.
23. Jiang, F., Pan, C., Zhang, Y., & Fang, Y. (2019). Cellulose graft copolymers toward strong thermoplastic elastomers via RAFT polymerization. *Applied Surface Science 480*, 162–171.
24. Meimoun, J., Wiatz, V., Saint-Loup, R., Parcq, J., Favrelle, A., & Bonnet, F. (2017). Modification of starch by graft copolymerization. *Starch–Stärke, 70*(1–2), 1600351.
25. Chauhan, N. P. S., Gholipourmalekabadi, M., & Mozafari, M. (2017). Fabrication of newly developed pectin–GeO2 nanocomposite using extreme biomimetics route and its antibacterial activities. *Journal of Macromolecular Science, Part A: Pure and Applied Chemistry, 54*(10), 655–661.
26. Sahariah, P., & Másson, M. (2017). Antimicrobial chitosan and chitosan derivatives: A review of the structure–Activity relationship. *Biomacromolecules, 18*(11), 3846–3868.
27. Argüelles-Monal, W., Lizardi-Mendoza, J., Fernández-Quiroz, D., Recillas-Mota, M., & Montiel-Herrera, M. (2018). Chitosan derivatives: Introducing new functionalities with a controlled molecular architecture for innovative materials. *Polymers, 10*(3), 342.
28. Marquis, M., Davy, J., Cathala, B., Fang, A., & Renard, D. (2015). Microfluidics assisted generation of innovative polysaccharide hydrogel microparticles. *Carbohydrate Polymers, 116*, 189–199.
29. Wong, T. W. (2011). Alginate graft copolymers and alginate-co-excipient physical mixture in oral drug delivery. *Journal of Pharmacy and Pharmacology, 63*(12), 1497–1512.
30. Odrobińska, J., Skonieczna, M., & Neugebauer, D. (2020). PEG graft polymer carriers of antioxidants: *In vitro* evaluation for transdermal delivery. *Pharmaceutics, 12*(12), 1178.
31. Inphonlek, S., Bureewong, N., Jarukumjorn, K., Chumsamrong, P., Ruksakulpiwat, C., & Ruksakulpiwat, Y. (2022). Preparation of poly (acrylic acid-co-acrylamide)-grafted deproteinized natural rubber and its effect on the properties of natural rubber/silica composites. *Polymers, 14*(21), 4602.
32. Lanzalaco, S., & Armelin, E. (2017). Poly (N-isopropylacrylamide) and copolymers: A review on recent progresses in biomedical applications. *Gels, 3*(4), 36.
33. Coudane, J., Van Den Berghe, H., Mouton, J., Garric, X., & Nottelet, B. (2022). Poly (lactic acid)-based graft copolymers: Syntheses strategies and improvement of properties

for biomedical and environmentally friendly applications: A review. *Molecules, 27*(13), 4135.
34. Coudane, J., Nottelet, B., Mouton, J., Garric, X., & Van Den Berghe, H. (2022). Poly (ε-caprolactone)-based graft copolymers: Synthesis methods and applications in the biomedical field: A review. *Molecules, 27*(21), 7339.
35. Neugebauer, D. (2007). Graft copolymers with poly (ethylene oxide) segments. *Polymer International, 56*(12), 1469–1498.
36. Zhang, Y., & Lam, Y. M. (2005). Poly (ethylene oxide)-b-poly (propylene oxide)-b-poly (ethylene oxide)-g-poly (vinyl pyrrolidone): Synthesis and characterization. *Journal of Colloid and Interface Science, 285*(1), 80–85.
37. Liu, B. (n.d.). Conjugated polymers for biological and biomedical applications | Wiley. Wiley.com. Retrieved from: https://www.wiley.com/en-cx/Conjugated+Polymers+for+Biological+and+Biomedical+Applications-p-9783527342730.
38. Hayward, A. S., Eissa, A. M., Maltman, D. J., Sano, N., Przyborski, S. A., & Cameron, N. R. (2013). Galactose-functionalized PolyHIPE scaffolds for use in routine three-dimensional culture of mammalian hepatocytes. *Biomacromolecules, 14*(12), 4271–4277.
39. Wang, L., Wu, X., Wang, X., Feng, Q., Pei, M., & Zhang, G. (2012). Synthesis and characterization of three novel conjugated polythiophene derivatives. *Designed Monomers and Polymers, 16*(4), 339–348.
40. Sand, A., & Vyas, A. (2020). Introductory chapter: Organic polymer–Graft copolymers. In *Organic polymer* (pp. 1–9). IntechOpen.
41. Purohit, P., Bhatt, A., Mittal, R. K., Abdellattif, M. H., & Farghaly, T. A. (2023). Polymer grafting and its chemical reactions. *Frontiers in Bioengineering and Biotechnology, 10, 1044927.*
42. Khedr, R. F. (2023). Radiation-grafting on polypropylene copolymer membranes for use in cadmium adsorption. *Polymers, 15*(3), 686.
43. Garnett, J. L., Jankiewicz, S. V., & Sangster, D. F. (1990). Mechanistic aspects of the acid and salt effect in radiation grafting. *Radiation Physics and Chemistry, 36*(4), 571–579. https://ui.adsabs.harvard.edu/abs/1990RaPC...36..571G/abstract (accessed on 17 September 2024).
44. Ashfaq, A., Clochard, M. C., Coqueret, X., Dispenza, C., Driscoll, M. S., & Ulański, P. (2020). Polymerization reactions and modifications of polymers by ionizing radiation. *Polymers, 12*(12), 2877.
45. Gupta, B., Anjum, N., & Gupta, A. P. (2000). Influence of solvents on radiation-induced graft copolymerization of acrylamide into polyethylene films. *Journal of Applied Polymer Science, 77*(6), 1401–1404.
46. Huang, R. Y. M. (1991). *Pervaporation Membrane Separation Processes*. Elsevier. Retrieved from: https://searchworks.stanford.edu/view/1733252 (accessed on 17 September 2024).
47. Ho, W. S. W., & Sirkar, K. K. (1992). *Membrane Handbook*. Springer.
48. Nishioka, N., Matsumoto, K., & Kosai, K. (1983). Homogeneous graft copolymerization of vinyl monomers onto cellulose in a dimethyl sulfoxide–paraformaldehyde solvent system II. Characterization of graft copolymers. *Polymer Journal, 15*(4). Retrieved from: https://www.semanticscholar.org/paper/Homogeneous-Graft-Copolymerization-of-Vinyl-onto-in-Nishioka-Matsumoto/523e55f1a233c6fe061342477869be49d119a560 (accessed on 17 September 2024).

49. Srinivasan, K. S. V., & Sudhakar, D. (1980). Gel permeation chromatography for characterization of graft copolymer. *Polymer Bulletin, 2*(5), 321–324. https://doi.org/10.1007/BF00299323.
50. Elsawy, M. A., Saad, G. R., & Elsabee, M. Z. (2011). Grafting of N-isopropyl acrylamide onto bacterial polyhydroxybutrate/hydroxyvalerate copolymers. *Polymer-Plastics Technology and Engineering, 50*(10), 1055–1063.
51. Thombare, N., Mishra, S., Siddiqui, M. Z., Jha, U., Singh, D., & Mahajan, G. R. (2018). Design and development of guar gum-based novel, superabsorbent and moisture-retaining hydrogels for agricultural applications. *Carbohydrate Polymers, 185*, 169–178.
52. Kim, B. S., & Mun, S. P. (2008). Effect of Ce4+ pretreatment on swelling properties of cellulosic super absorbents. *Polymers for Advanced Technologies, 20*(12), 899–906.
53. Kubota, H., Fukushima, Y., & Kuwabara, S. (1997). Factors affecting liquid-phase photo grafting of acrylic acid on cellulose and its derivatives. *European Polymer Journal, 33*(1), 67–71.
54. Toledano-Thompson, T., Loria-Bastarrachea, M. I., & Aguilar-Vega, M. J. (2005) Characterization of henequen cellulose microfibers treated with an epoxide and grafted with poly (acrylic acid). *Carbohydrate Polymers, 62*(1), 48–56. https://agris.fao.org/agris-search/search.do?recordID=US201301044964 (accessed on 17 September 2024).
55. Kim, S., Yu, H., Jeong, C. H., Choi, E., & Ahn, S. (2021). Composition analysis of grafted and its blend copolymers using a combination of spectroscopy and nitrogen element analysis. *Bulletin of the Korean Chemical Society, 42*(9), 1251–1258. https://doi.org/10.1002/bkcs.12227.
56. Mahmoodi, N. O., Ahmadi, A., Yazdani Nyaki, H., Taherpour Nahzomi, H., & Panahi Kokhdan, E. (2022). Efficient synthesis of new azo-sulfonamide derivatives and investigation of their molecular docking and cytotoxicity results. *Arabian Journal of Chemistry, 15*(12), 104383.
57. Hiwatari, K., Serizawa, T., Kishida, A., & Akashi, M. (2001). Graft copolymers having hydrophobic backbone and hydrophilic branches XXXI. A DSC study of internal structures and thermophysical properties of core-corona polymeric nanospheres. *Polymer Journal, 33*(5), 424–428.
58. Ferrero, M. C., Velasco, M. V., Ford, J. L., Rajabi-Siahboomi, A. R., Muñoz, A., & Jiménez-Castellanos, M. R. (1999). Determination of the glass transition temperatures of some new methyl methacrylate copolymers using modulated temperature differential scanning calorimetry (MTDSC). *Pharmaceutical Research, 16*(9), 1464–1469.
59. Adjuik, T. A., Nokes, S. E., & Montross, M. D. (2023). Grafting of modified cellulose nanofibrils onto chitosan-polyethylene oxide films. *Journal of Applied Polymer Science, 140*(12), e53655.
60. Najer, A., Wu, D., Nussbaumer, M. G., Schwertz, G., Schwab, A., & Witschel, M. C. (2016). An amphiphilic graft copolymer-based nanoparticle platform for reduction-responsive anticancer and antimalarial drug delivery. *Nanoscale, 8*(31), 14858–14869.
61. Van Camp, W., Germonpré, V., Mespouille, L., Dubois, P., Goethals, E. J., & Du Prez, F. E. (2007). New poly (acrylic acid)-containing segmented copolymer structures by combination of chemistry and atom transfer radical polymerization. *Reactive and Functional Polymers, 67*(11), 1168–1180.
62. Zhao, Y. (2017). Asymmetrical polymer vesicles for drug delivery and other applications. *Frontiers in Pharmacology, 8*(1), 374.

63. Tanner, P., Baumann, P., Enea, R., Onaca, O., Palivan, C., & Meier, W. (2011). Polymeric vesicles: From drug carriers to nanoreactors and artificial organelles. *Accounts of Chemical Research, 44*(10), 1039–1049.
64. Wang, Y., Pai, J. H., Lai, H. H., Sims, C. E., Bachman, M., & Li, G. P. (2007). Surface graft polymerization of SU-8 for bio-MEMS applications. *Journal of Micromechanics and Microengineering, 17*(7), 1371–1380.
65. Julthongpiput, D. (2003). *Design of Nanocomposite Polymer Coatings for MEMS Applications* (Doctoral dissertation, Iowa State University.
66. Smirnov, M. A., Tarasova, E. V., Vorobiov, V. K., Kasatkin, I. A., Mikli, V., & Sokolova, M. P. (2018). Electroconductive fibrous mat prepared by electrospinning of polyacrylamide-g-polyaniline copolymers as electrode material for supercapacitors. *Journal of Materials Science, 54*(6), 4859–4873.
67. Wang, J., & Higashihara, T. (2013). Synthesis of all-conjugated donor–Acceptor block copolymers and their application in all-polymer solar cells. *Polymer Chemistry, 4*(22), 5518–5526.
68. Jarosz, T., Gebka, K., Kepska, K., Lapkowski, M., Ledwon, P., & Nitschke, P. (2018). Investigation of the effects of non-conjugated co-grafts on the Spectro electrochemical and photovoltaic properties of novel conjugated graft copolymers based on poly (3-hexylthiophene). *Polymers, 10*(10), 1064.
69. Antony, S., Anju, T. R., & Thomas, B. (2022). Nature-inspired biomimetic polymeric materials and their applications. In *Handbook of Biopolymers* (Vol. 10, pp. 1–31).
70. Ganewatta, M. S., Wang, Z., & Tang, C. (2021). Chemical syntheses of bioinspired and biomimetic polymers toward biobased materials. *Nature Reviews Chemistry, 5*(11), 753–772.
71. Coggio, W., Chandler, J. M., & Thompson-Barclay, S. L. (2023). WIPO–Search International and National Patent Collections. *Patentscope*. Retrieved from: https://patentscope.wipo.int/search/en/detail.jsf?docId=US394061095&_cid=P10-LG9JFR-13488-1 (accessed on 17 September 2024).
72. Donavan, L., Jeremy, G., Wenjun, W., & Jeffery, S. (2023). WIPO–Search International and National Patent Collections. *Patentscope*. Retrieved from: https://patentscope.wipo.int/search/en/detail.jsf?docId=EP392329400&_cid=P10-LG9KWL-41474-1 (accessed on 17 September 2024).
73. Hui, S. H., Dong, Y., & Jiyang, Z. (2022). WIPO–Search International and National Patent Collections. *Patentscope*. Retrieved from: https://patentscope.wipo.int/search/en/detail.jsf?docId=CN369012341&_cid=P10-LG9J8N-10027-1 (accessed on 17 September 2024).
74. Brown, A., George, W., Richez, A., Dingwall, A. C., & Von Hatten, X. (2022). WIPO–Search International and National Patent Collections. *Patentscope*. Retrieved from: https://patentscope.wipo.int/search/en/detail.jsf?docId=JP373480903&_cid=P10-LG9JXZ-21817-1 (accessed on 17 September 2024).
75. Xing, W. (2021). WIPO–Search International and National Patent Collections. *Patentscope*. Retrieved from: https://patentscope.wipo.int/search/en/detail.jsf?docId=CN345730292&_cid=P10-LG9L90-48670-1 (accessed on 17 September 2024).
76. Chang, Y. E. (2021). WIPO–Search International and National Patent Collections. *Patentscope*. Retrieved from: https://patentscope.wipo.int/search/en/detail.

jsf?docId=TH342838197&_cid=P10-LG9LCM-50525-1 (accessed on 17 September 2024).
77. Terruzzi, C. (2020). Compositions for the treatment of epithelial lesions. *European Patent Office*. Retrieved from: https://data.epo.org/gpi/EP3811960A1-compositions-for-the-treatment-of-epithelial-lesions (accessed on 17 September 2024).
78. Ruo Peng, L., Zhi Ya, Z., & Cheng Wei, X. (2020). Protective coating and method of use thereof. *WIPO*. Retrieved from: https://patentscope.wipo.int/search/en/detail.jsf?docId=WO2019062083 (accessed on 17 September 2024).
79. Heremel-Davidock, T., & Coughlin, E. B. (2019). WIPO–Search International and National Patent Collections. *Patentscope*. Retrieved from: https://patentscope.wipo.int/search/en/detail.jsf?docId=MX290133163&_cid=P10-LG9JQB-18411-1 (accessed on 17 September 2024).
80. Li, J., Bai, X., Liu, X., Ren, Q., & Wang, C. (2018). Preparation method of polyethylene glycol grafted polymethyl methacrylate amphiphilic graft copolymer. *Google Patents*. Retrieved from: https://patents.google.com/patent/CN107903348B/en (accessed on 17 September 2024).

CHAPTER 3

Polymer Derivatives for Drug Delivery Application

ANAS ISLAM,[1] USAMA AHMAD,[1] SYED FAREHA,[2] and MOHD. MUAZZAM KHAN[1]

[1]*Faculty of Pharmacy, Integral University, Lucknow, Uttar Pradesh, India*

[2]*Department of Bioengineering, Integral University, Lucknow, Uttar Pradesh, India*

ABSTRACT

The field of drug delivery has undergone significant advancements over the past few decades, and the progress of innovative drug delivery systems (IDDSs) has turned out to be a crucial aspect of the pharmaceutical industry. One such system that has gained significant consideration in recent years is the use of graft copolymers of polysaccharides as drug delivery vehicles. Polysaccharides, being biocompatible and biodegradable, have emerged as attractive polymers for the design of DDSs. The combination of polysaccharides with synthetic polymers to form graft copolymers enhances the properties of both polymers and results in the progress of novel DDSs. Current developments in the design and synthesis of graft copolymers of polysaccharides have shown promising results in many applications, including the controlled release of drugs, improved stability and bioavailability of drugs, and targeted drug delivery. Despite the promising results, the use of graft copolymers of polysaccharides as DDSs still faces certain challenges. One of the major challenges is the lack of stability of the copolymers in biological environments, leading to rapid degradation and reduced drug release efficacy. Additionally, the synthesis of graft copolymers can be complex and time-consuming, and there is a need for more efficient and cost-effective synthetic methods. Furthermore,

Graft Copolymers of Polysaccharides for Drug Delivery Systems.
Rishabha Malviya and Sonali Sundram (Eds.)

the understanding of the interactions between the drug and the copolymer and the mechanisms of drug release is still limited, and further research is required in this area. In conclusion, the use of graft copolymers of polysaccharides as DDSs represents a promising area of research with potential applications in various therapeutic areas. Despite the challenges, the current developments in the field have shown significant potential, and there is a growing interest in the design and progress of novel graft copolymer-based DDSs. Further research is needed to overcome the limitations and to fully realize the potential of these systems in the development of effective and safe DDSs.

3.1 INTRODUCTION

A drug delivery system (DDS) is a formulation or a piece of equipment that enables the safe and controlled administration of an advantageous material into the body. This expands the medicine's efficacy and safety by controlling the timing, rate, and location of drug release. The medicinal product's administration, the release of its active chemicals, and their transportation through biological membranes to the place of action are all included in this process. The term "therapeutic substance" encompasses agents like gene therapy, which induces the production of active therapeutic agents within the body. A drug formulation or a delivery device may take the form of a DDS, which serves as the interface between the patient and the drug, essentially. This distinction is important, as it determines whether regulatory agencies governing drug control will oversee the delivery system. For example, a device will be rigorously controlled as a device if it is inserted into the body for therapeutic purposes unrelated to medication delivery or if a drug is included in the device to prevent issues with the device. The classification of a particular system as a drug or device is determined on an individual basis, as there is a broad range between drugs and devices [1]. The process of developing a new drug molecule is both time-consuming and costly. As a result, scientists have worked to improve the safety and effectiveness of already available medications using a range of techniques, such as dosage titration, therapeutic drug monitoring, and individualizing drug therapy. Additionally, a lot of research has been done on the controlled delivery of medications at a slow or targeted rate because it is an especially promising strategy [2–4]. Inhalation anesthetics, sedative/hypnotic medications, muscle relaxants, and potent opioid analgesics have all been investigated extensively in recent years in both animal and human trials to determine the pharmacokinetic (PK) and pharmacodynamic principles governing their action and disposition.

These investigations recommend that buccal, cutaneous, and nasal mucous membranes may provide additional pathways for the administration of analgesics and anesthetics. Since then, several different controlled-release technologies (CRTs) have been developed, including transdermal and transmucosal controlled-release delivery systems, implanted drug-delivery devices, buccal and nasal aerosol sprays, encapsulated cells, drug-impregnated lozenges, nasal and buccal aerosol sprays, and oral soft gels. The creation of these new tools, ideas, and methods is motivated by a variety of circumstances. These technologies have the potential to solve a number of issues with conventional medication delivery procedures. The development of new drugs has become more expensive due to rising R&D costs, diversification of investment opportunities for pharmaceutical companies, a decline in the number of organizations accompanying pharmaceutical studies, and the eroding of effective patent life. These innovations may provide a more cost-effective alternative to this process. Currently, it takes an estimated 10 years and more than $120 million to develop a novel medication through exploration, clinical trials, progress, and regulatory approval [5–7].

Effective delivery strategies have played a critical role in converting promising therapeutics into successful therapies [8–11]. Delivery tactics and technology have quickly changed to match the shifting demands of medication delivery as the therapeutic environment has changed. Small-molecule medications were the main category of therapies a few decades ago. Small molecules' physicochemical characteristics, which have a big impact on medication bioavailability, substantially determine how well they are delivered. Hence, early delivery efforts focused on increasing drug solubility, regulating release, boosting activity, and altering PKs [12, 13]. Nucleic acids, monoclonal antibodies (mAbs), proteins, peptides, and living cells are examples of the new generations of medicines that have evolved over time and offered novel therapeutic activities. But with greater power comes greater complexity, particularly in terms of stability (primarily for peptides and proteins), survival and growth (for living cells), and intracellular transport requirements (primarily for nucleic acids). Therefore, to address these issues, drug delivery methods must evolve [14]. By delivering and releasing the medication to the target place in the body while minimizing off-target buildup, drug delivery ultimately aims to maximize therapeutic efficacy. Controlling medication PKs, lowering drug toxicity, boosting drug accumulation at the target location, and enhancing patient compliance and acceptability can all help with this. The revolution in delivery technology and methods has been sparked by the recognition of various delivery difficulties that are unique to each therapeutic class [14].

Graft polymerization is the process of adding side chains made of monomers to the primary polymer chain, or backbone, by covalent bonding and polymerization. Incorporating different functional groups into a polymer can be achieved effectively through the process of grafting. The resultant polymer is classified as a graft polymer or graft copolymer because it contains grafted side chains that are physically different from the main chain and at least two diverse types of monomer units. A grafted copolymer's graft chains may be homopolymers or copolymers since the monomer to be grafted may be of one or more kinds [15]. "Grafting to" and "grafting from" are the two primary approaches to grafting. "Grafting to" methods involve the reaction of functionalized monomers with the backbone polymer to create the grafted polymer. On the other hand, "grafting from" is accomplished by treating a substrate with an appropriate technique to generate immobilized initiators, followed by polymerization. This technique also enables the production of high grafting-density polymers [16]. Enzymatic grafting, plasma-induced procedures, photoirradiation, ionizing radiation, chemical therapy, photochemical treatment, and ionizing radiation are some of the methods that can initiate graft copolymerization. Grafted polymers are highly versatile, as they can be customized for specific applications by selecting the appropriate backbone and monomers to be grafted [15].

A significant advantage of graft copolymers is their ability to tune their amphiphilicity, enabling them to adjust their hydrophilic-hydrophobic balance based on the surrounding environment. This characteristic allows them to self-assemble into a wide variety of nanostructures, such as vesicles, hydrogels, micelles, and so forth. These nanostructures can encapsulate or conjugate drugs and protect them from degradation and premature release. Additionally, graft copolymers can retort to various spurs, such as pH, temperature, redox potential, light, and so forth, that can induce drug release at specific sites and times as desired [17–19]. Several advantages of graft copolymers for DDSs are demonstrated in Figure 3.1 (Table 3.1).

3.2 GRAFT COPOLYMERS USED IN DRUG DELIVERY

Graft copolymers find applications in drug delivery due to their ability to increase solubility, stability, biocompatibility, and drug targeting [20]. Here are some examples of graft copolymers used in drug delivery:

3.2.1 POLYETHYLENE GLYCOL (PEG) GRAFT COPOLYMERS

Polyethylene glycol (PEG) is a synthetic polymer that is both hydrophilic and biocompatible. It is widely used to modify drugs, proteins, peptides,

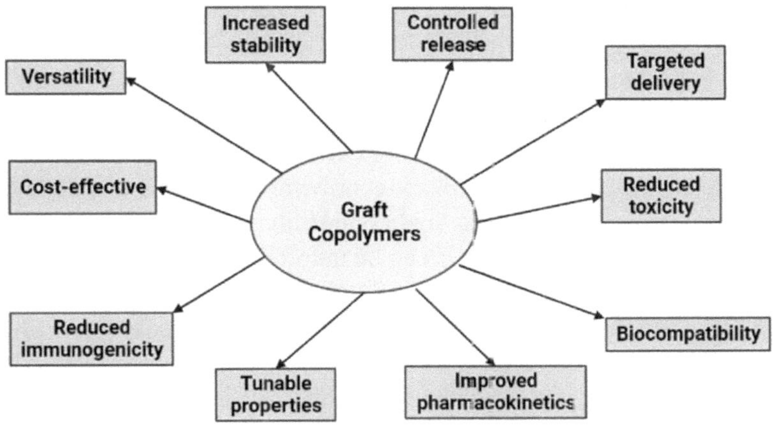

FIGURE 3.1 Advantages of graft copolymers for drug delivery systems.

TABLE 3.1 Comparison of Polysaccharide-Based Graft Copolymers and Other Nanoparticle-Based Drug Delivery Systems

Parameter	Polysaccharide-Based Graft Copolymers	Other Nanoparticle-Based Drug Delivery Systems
Biodegradability	High biodegradability due to natural origin	Varies depending on the material used; some are biodegradable, others are not
Biocompatibility	Generally high biocompatibility due to natural origin	Varies depending on the material used; some are biocompatible, while others may have issues
Drug Loading Capacity	Limited drug loading capacity due to hydrophilic nature	Varies; can have high drug loading capacity for lipophilic drugs
Targeted Drug Delivery	May require additional modifications for targeted drug delivery	Targeted drug delivery can be achieved through surface modifications
Stimuli-Responsive Drug Release	Varies depending on the graft component	Varies depending on the material used and its properties
Mucoadhesive Properties	Some polysaccharide-based graft copolymers, like chitosan, have mucoadhesive properties	Generally, lack mucoadhesive properties unless surface-modified
Solubility Enhancement	Often, good solubility enhancement due to the hydrophilic nature of polysaccharides	Varies depending on material used
Immunogenicity	Generally low immunogenicity due to natural origin	Varies depending on the material used; some materials may have immunogenicity issues
Ease of Synthesis and Modification	Moderate to difficult due to complex structure and hydrophilic nature	Varies depending on the material used; some are easier to synthesize and modify

and nanocarriers to enhance their solubility, stability, and biocompatibility. PEG can be grafted onto other polymers, creating block copolymers with varying properties and functionalities. An example of a block copolymer is polyethylenimine-graft-poly(ethylene glycol) (PEI-g-PEG), which is cationic and thermoresponsive, forming complexes with negatively charged oligonucleotides or ribozymes. The block copolymer can serve as a gene therapy DDS by safeguarding, carrying, and controlling the release of nucleic acids. The PEI-g-PEG block copolymer can be tailored into diverse structures, such as nanoparticles, micelles, or hydrogels, based on the application. *In vitro* and *in vivo* research has verified the biodegradability and safety of PEI-g-PEG block copolymer, demonstrating its potential for biomedical purposes [21–23].

The FDA has given the go-ahead for PEG to be used as a base or carrier in a variety of goods, including meals, cosmetics, and medications that are administered orally, topically, recto-rectally, or nasally. PEG is totally eliminated from the body and has a low degree of toxicity; it does so either through the kidneys for PEGs under 30 kDa or through the stool for PEGs above 20 kDa. PEG is not immunogenic, and rabbits only develop antibodies to PEG when PEG is coupled with proteins that are highly immunogenic [24, 25]. No reports of the development of PEG antibodies during the usual clinical delivery of pegylated proteins have been made so far. Animals have, however, developed antibodies against PEG as a result of being injected with PEG-protein conjugates under rigorous experimental settings [26]. PEG is synthesized by linking repeating units of ethylene glycol to create polymers of varying molecular masses, which can have linear or branched shapes. The chosen medication is linked chemically to the PEG architectures by the method of pegylation. Investigations on PEG when in a solution demonstrate that every ethylene glycol component tightly interacts with two or three water molecules. As a result of water molecule binding, pegylated substances exhibit the behavior of being 5 to 10 times larger than a soluble protein having a similar molecular mass. To verify this, size exclusion chromatography and gel electrophoresis were employed [27]. PEG and the accompanying water molecules shield the linked medicine from enzymatic degradation, quick renal clearance, and association with cell surface proteins, thus decreasing unfavorable immunological reactions. Pegylated medications are also more resistant to changes in pH and temperature than their unpegylated equivalents [28]. Pegylation confers numerous qualities on medications as a consequence, including sustained blood levels that increase efficacy, decreased frequency of adverse reactions, a longer shelf life, and better patient convenience, which are anticipated to have positive clinical effects [29]. An extension in the half-life of biological systems can compensate for the negative impact that pegylation could have on the *in vitro* activity of proteins. Pegylation has the potential to change the bioactivity of polypeptides by influencing the ability of therapeutic proteins to bind with cellular receptors [29]. In order to fully utilize the benefits of pegylation in polypeptide medication, it is crucial to create a solid connection between the PEG polymer and the medication. This usually involves activating the PEG polymer chemically before it can bind with the medication [30]. Several chemical alterations are utilized to produce a PEG derivative that can attach to a particular target molecule. These modifications lead to the production of an active PEG derivative that includes a functional group like an active carbonate, aldehyde, active ester, or tresylate. Subsequently, covalent attachment is employed to connect this derivative to a reactive group on the polypeptide drug. The N-terminal amino acid groups of various amino acids or the amino groups of lysine are frequently used as reactive sites on polypeptides for joining PEG polymers [31]. The biological action of the medication coupled to PEG polymers can be affected by changes in their size, structure, and molecular weight. A polypeptide's half-life is typically extended, its renal clearance is decreased, and its biological activity is increased when it is pegylated.

3.2.2 POLYVINYL ALCOHOL (PVA) GRAFT COPOLYMERS

For a long time now, biomedicine has made use of PVA, which is a biodegradable synthetic polymer with a semi-crystalline structure. To enhance its properties and functionality, PVA can undergo polymer grafting with other polymers. One instance of such modification is the thermoresponsive material, poly(vinyl alcohol)-graft-poly(N,N-diethylacrylamide) (PVA-g-PDEAAm) copolymer. The PVA-g-PDEAAm copolymer is a perfect DDS for the regulated release of therapeutic drugs in the treatment of cancer because it may alter its swelling behavior in response to temperature. The copolymer can be fabricated into different forms, including hydrogels, microparticles, or nanoparticles, based on the intended application. Biocompatibility and non-toxicity of PVA-g-PDEAAm copolymer have been confirmed in *in-vitro* and *in-vivo* studies, indicating that it has promising potential for use in biomedical applications [32, 33].

The biodegradable semicrystalline synthetic polymer, PVA, is frequently employed by the pharmaceutical sector to form solid dispersions that elevate drug solubility and prevent crystal transformation. Furthermore, PVA operates as a granulating agent that hinders nucleation, delays crystal formation, and absorbs excess water to eliminate it [34, 35]. Since the 1970s, PVA has also been utilized as an embolic agent, with its current use in intravascular applications being predominantly in the form of microparticles [36]. PVA and its copolymers have recently been used to control medication distribution since they work well with both hydrophilic and hydrophobic medicines. Due to the high median lethal doses (LD50) of PVA in rats, it is also thought to be non-toxic when supplied orally. PVA is a frequent excipient in medications delivered by ocular, vaginal, and transdermal routes [34, 37].

PVA is a versatile material that finds applications in diverse fields, including biotechnology, hydrogel formation, drug delivery, biosensors, and biomedical devices, owing to its promising properties [38–42]. Incorporating copolymer groups into PVA and converting them into different polymers expands its range of potential applications [43, 44] Numerous fields have paid close attention to the innovative design structures of graft and block copolymers that are enhanced with copolymer groups [45]. Graft polymers exhibit diverse structures, molecular weights, and functionality, which contribute to their high performance and enable the controllable modification of polymer material properties [46, 47]. The manipulation of polymer material characteristics is made possible through the use of controlled radical polymerization methods, such as ATRP, which provide precise control. Numerous studies have been conducted on controlled radical polymerization to advance this field [48–52]. Due to its distinct benefits over live polymerization processes, such as stability and popularity, RAFT polymerization has attracted a lot of attention since it was first discovered [53].

3.2.3 POLY (N-VINYL PYRROLIDONE) (PVP) GRAFT COPOLYMERS

Poly (N-vinyl pyrrolidone) (PVP) is a water-soluble polymer that can merge with other biodegradable polymers, including polylactic acid (PLA), to develop amphiphilic graft copolymers. These copolymers can take on various forms, such as microparticles, nanoparticles, hydrogels, or films, and can be utilized to control drug release. PVP segments within the copolymers are beneficial in improving drug solubility, stability, biocompatibility, and bioavailability, while the biodegradable segments allow for adjustable degradation rates and drug release kinetics. Researchers have studied PVP-based graft copolymers for various drug delivery applications, such as anticancer therapy, wound healing, gene delivery, and inhibiting COVID-19 [54–56].

Polyvidone or povidone, also known as PVP, is a biodegradable water-soluble polymer made from the N-vinylpyrrolidone (NVP) monomer. Apart from being hydrophilic, PVP displays exceptional solubility in solvents with varying polarities and has the capacity to bind effectively, as well as stabilize suspensions and emulsions [57]. The characteristics and operation of graft copolymers based on PVP can be altered to match various applications by regulating the molecular weight, composition, and grafting density of the side chains and backbone. [58–62]. PVP displays unique physical and chemical characteristics, such as chemical inertness, colorlessness, temperature resistance, and pH stability. These properties render PVP an attractive polymer for the production of pharmaceutical formulations. PVP enables the controlled liberation of medications, boosts the bioavailability of drugs that are poorly soluble in water, provides protection to active compounds from outside elements (such as pH, temperature, and oxygen), and masks disagreeable odors and tastes. Researchers have incorporated diverse, active compounds into PVP microparticles and nanoparticles using various techniques, ranging from conventional methods (e.g., spray drying) to innovative techniques (e.g., supercritical fluids-assisted methods). PVP's spinnability allows for the creation of PVP fibers laden with different active ingredients, while PVP-based hydrogels and oral tablets have been proposed in some studies. Additionally, PVP displays excellent film-forming abilities. The distinctive and adaptable characteristics of PVP render it a compelling polymer for developing pharmaceutical preparations, and its application in up-and-coming fields of biomedicine and pharmacy may furnish long-awaited answers for the pharmaceutical sector [56].

3.2.4 POLY (ETHYLENE OXIDE) (PEO) GRAFT COPOLYMERS

Poly (ethylene oxide) (PEO) is a hydrophilic polymer that has the ability to produce amphiphilic graft copolymers with other hydrophobic polymers such as poly(ε-caprolactone) (PCL), poly (propylene oxide) (PPO), or poly (ester amide) (PEA). In an aqueous solution, micelles or nanoparticles can be naturally formed by these graft copolymers that can encapsulate various active pharmaceutical ingredients (APIs) for controlled drug delivery. The PEO segments in the copolymers contribute to improving the biocompatibility, solubility, and stealthiness of the nanocarriers, whereas the hydrophobic segments provide stability, biodegradability, and drug loading capacity [20, 63].

A hydrophilic polymer, polyethylene oxide interacts with water molecules, giving it this characteristic. Polyethylene oxide (PEO), also referred to as Polyox, is a non-ionic homopolymer of ethylene oxide. White powders composed of PEO with different molecular weights, spanning from 100,000 to 8 million, possess an average particle size of roughly 150 mm. PEO acts as a matrix material, and the polymer's high molecular weight versions can reduce the release rate of both insoluble and soluble medications from matrix tablets created by direct compression. Furthermore, PEO can function as a mucoadhesive polymer by adhering to mucosal surfaces inside the body. Polyox water-soluble resins are employed for various pharmaceutical purposes, including transdermal DDSs, instant release and gastro-retentive dosage forms, tablet coating, and tablet binding [64]. Polyethylene oxide showcases water retention and film-forming features, which make it practical for a variety of uses. It is immensely dissolvable in water and has minimal toxicity [65].

Polyox is among the swiftest water-soluble polymers for hydration in pharmaceutical systems. It expeditiously generates hydrogels that stimulate and regulate the liberation of active components. The singular swelling characteristics of Polyox, combined with its controlled dissolution rate, render it an optimal choice for timed-release formulations. Polyox is non-poisonous and has been sanctioned by the FDA for diverse food and drug applications. Moreover, polyox aqueous solutions can decompose environmentally through oxidation and aerobic biodegradation. In animal research, Polyox was shown to possess extremely low levels of toxicity through all routes of exposure, with the largest practical oral dose for rats being around 2 g/kg of body weight [64].

3.2.5 DEXTRAN-POLY (ACRYLIC ACID) GRAFT COPOLYMER

Dextran is a type of polysaccharide that is soluble in water and primarily consists of D-glucopyranose residues that are linked by 1,6 bonds. Its low tissue toxicity and great biodegradability are what give it its distinctive physicochemical characteristics and biocompatibility. Moreover, the presence of numerous active hydroxyl groups allows for easy modification and introduction of various functional groups onto its skeletal structure. Dextran is frequently recognized as a viable matrix system for drug administration or controlled release of bioactive substances as a result of these characteristics [65–68]. Due to its susceptibility to stimuli like temperature and pH level, dextran has favorable properties but unsatisfactory release properties. The existence of carboxyl groups in the synthesized dextran-based nanoparticles is responsible for their strong pH sensitivity, according to research conducted on this topic [69]. Researchers reported a novel carboxylmethyl dextran hydrogel membrane, which exhibits pH- and ionic-strength-sensitivity. The membrane's swelling behavior is pH-dependent and is attributed to the presence of acidic groups [70]. A graft copolymer consisting of dextran and poly (acrylic acid) (PAA) has been developed, which can form pH-sensitive hydrogels. These hydrogels exhibit reversible swelling or shrinkage depending on the environmental pH and can effectively release drugs in a controlled manner.

3.2.6 CHITOSAN-G-POLY (ETHYLENE GLYCOL) GRAFT COPOLYMERS

Chitosan is a polycationic biopolymer that shares structural characteristics with glycosaminoglycans (GAGs). It is derived from chitin, the main element of the exoskeleton of crustaceans such as shrimp, through alkaline deacetylation [71]. The biodegradability, biocompatibility, mucoadhesivity, pH sensitivity, low immunogenicity, and low toxicity are just a few of the intriguing biopharmaceutical characteristics of chitosan. These advantageous properties have lately increased interest in chitosan and its derivatives as delivery methods for specific medication administration to tumors [72–74]. Because its primary amine groups are located in the C-2 position of its monomeric units and its hydroxyl groups are located in the C-6 position, chitosan is very adaptable to specific chemical alterations, resulting in a wide range of derivatives. Chemically altered chitosan derivatives are very useful for developing medication carriers that specifically target tumors. Chitosan is one of the most popular biopolymers for NP synthesis because of its distinctive structural characteristics. It has the ability to crosslink with different types of crosslinking materials, including sodium tripolyphosphate (TPP), genipin, and glutaraldehyde, to build a strong network that can trap drug molecules inside nanoparticles. Chitosan-based nanoparticles have longer blood circulation durations and are less likely to be absorbed by the reticuloendothelial system (RES) because of their smaller size. Moreover, they can aim for cancerous growth locations by utilizing the enhanced permeability and retention (EPR) phenomenon that occurs as a result of the flawed and incoherent vascular system of tumor tissues [75]. Chitosan and poly(ethylene glycol) (PEG) graft copolymers can create micelles that can improve both the solubility and the stability of hydrophobic medicines. Additionally, by taking advantage of the improved permeation and retention (EPR) effect, these micelles can specifically target tumor cells.

3.2.7 STARCH-G-POLY (N-ISOPROPYLACRYLAMIDE) GRAFT COPOLYMERS

The two primary components of starch, amylose, and amylopectin, which have different chain structures, are complicated carbohydrates. Amylose is made up of α-(1–4)-d-glycoside linkages that hold linear chains of anhydrous glucose units together. Amylose has a molecular weight of fewer than a million on average. Amylopectin, on the other hand, has a large branching structure with α-(1–6) connections, giving it a molar mass that can reach hundreds of millions [76–79]. Starch is an appealing substance for various applications due to its non-toxic, biodegradable, and biocompatible nature as a polymer [80, 81]. Starch is an extremely adaptable polymer that has drawn a lot of interest as a means of medication administration. Additionally, in recent years, SNPs, a type of nanoparticle, have been investigated as potential drug carriers. Nanoparticles can facilitate the prolonged administration of numerous chemicals to diverse regions of the body. Compared to microparticles, nanoparticles exhibit greater intracellular absorption. Additionally, their diminutive size and maneuverability enable them to reach a wide range of biological tissues. However, it should be noted that because of the size of their capillaries, organs like the spleen have a tendency to accumulate particles larger than 230 nm [82]. In-depth research has been done on the use of starch as a drug delivery vehicle for a variety of medications. Transdermal and tumor-targeted drug delivery are particularly appealing. Drugs, including aspirin, estrone, and phenethylamines, have all been transported via it. Chemically modified starch nanoparticles (SNPs) have been implemented in sustained DDSs. For controlled contramid liberation, crosslinked high amylose starch has been utilized as a matrix. Propylated SNPs carrying drug loads have indicated enhanced penetration through human skin. When compared to free 5-Fu, dialdehyde starch nanoparticles (DASNPs) associated with 5-fluorouracil have displayed improved inhibition against breast cancer cells (MCF-7) *in vitro* [83]. A graft copolymer composed of starch and poly (N-isopropylacrylamide). It can form temperature-sensitive hydrogels that can swell or shrink depending on the environmental temperature, and can release drugs in a pulsatile manner.

3.2.8 OTHER GRAFT COPOLYMERS USED IN DRUG DELIVERY

- **Poly (ε-Caprolactone) Graft Copolymer:** For the graft copolymer amphiphilic to produce steady micelles in aqueous solutions, a low critical micelle concentration (CMC) is needed. Because it can both enclose hydrophobic drugs and associate pH-sensitive chemotherapeutics to target tumors, it is a potential option for effective dual-drug delivery. This copolymer can be synthesized using controlled RAFT polymerization and click chemistry and is capable of degrading and releasing drugs when exposed to lipase [84].
- **Dextran-β-Cyclodextrin/Ferrocene-Poly (lactide) (Dex-β-CD/ Fc-PLA)**: A novel noncovalent graft copolymer consisting of β-cyclodextrin (β-CD) grafted dextran (Dex-β-CD) and ferrocene (Fc) terminated poly(lactide) (Fc-PLA). The copolymer can self-assemble into spherical nanoparticles that exhibit triple stimuli-responsiveness to

changes in pH, oxidation, and electric fields. These nanoparticles have the ability to encapsulate hydrophobic anticancer drugs and free them in a triggered style under simulated oxidizing and acidic conditions [85].

- **Polyurethane Graft on Linear Dextrin**: Because of its biocompatibility and biodegradability, this copolymer is an appropriate substance for tissue engineering since it can facilitate cell adhesion and propagation. The porous scaffolds constructed from this copolymer can furnish a three-dimensional milieu for cell growth, tissue regeneration, and drug administration. The mechanical characteristics and hydrophilicity of the scaffolds can be regulated by monitoring the extent of crosslinking, grafting density, and dextran chain length. This copolymer has the potential to be used in the restoration and regeneration of various tissues, including bone, cartilage, and skin [86].

General overview of the advantages and disadvantages of the above listed graft copolymers in DDSs are summarized in Table 3.2.

3.3 INNOVATIVE DRUG DELIVERY SYSTEMS USING GRAFT COPOLYMERS

IDDSs using graft copolymers are a promising approach for enhancing the efficiency and safety of therapeutic agents. Graft copolymers are polymers that have one or more side chains attached to the main chain. They can be tailored to have different properties such as amphiphilicity, biodegradability, bio-reducibility, stimuli-responsiveness, and targeting ability. These characteristics may facilitate the formation of self-arranged compositions by the graft copolymers, such as micelles, nanoparticles, or conjugates, which can sequester or associate with drugs and transport them to designated areas within the body. Some examples of IDDSs using graft copolymers are:

3.3.1 DESIGNING GRAFT COPOLYMERS FOR DUAL DRUG DELIVERY WITH PH-SENSITIVE DRUG CONJUGATION AND VARIABLE AMPHIPHILICITY

Researchers have developed copolymers as a means of transporting both hydrophobic and chemotherapeutic chemicals to cancer cells, offering an effective cancer therapy. These copolymers, specifically poly(-caprolactone)-graft-(poly-N-(2-hydroxypropyl) methacrylamide) (PCL-graft-pHPMA), can create micelles that are highly stable, with a low CMC and a high drug

TABLE 3.2 General Overview of the Advantages and Disadvantages of Graft Copolymers in Drug Delivery Systems

Material	Advantages	Disadvantages
Polyethylene Glycol (PEG) Graft Copolymers	• Biocompatible • Hydrophilic • Enhances solubility • Reduces immunogenicity	• Possibility of PEG-specific antibodies • Limited drug loading capacity
Polyvinyl Alcohol (PVA) Graft Copolymers	• Biocompatible • Hydrophilic • Good mechanical properties • Low toxicity	• Poor solubility in some solvents • Limited drug loading capacity
Poly(N-Vinyl Pyrrolidone) (PVP) Graft Copolymers	• Biocompatible • Hydrophilic • Enhances solubility • Excellent film-forming properties	• Limited drug loading capacity • Potential for hydrolytic degradation
Poly(Ethylene Oxide) (PEO) Graft Copolymers	• Biocompatible • Hydrophilic • Enhances solubility • Reduces protein adsorption	• Limited drug loading capacity • Possibility of oxidative degradation
Dextran-Poly(Acrylic Acid) Graft Copolymer	• Biocompatible • pH-responsive • Controlled drug release • Good water solubility	• Difficult to control grafting – swelling and degradation issues

TABLE 3.2 *(Continued)*

Material	Advantages	Disadvantages
Chitosan-g-Poly(Ethylene Glycol) Graft Copolymers	• Biocompatible • Hydrophilic • Mucoadhesive • Controlled drug release • Biodegradable	• Poor solubility in some solvents • Limited drug loading capacity
Starch-g-Poly(N-Isopropylacrylamide) Graft Copolymers	• Biocompatible • Thermoresponsive • Controlled drug release – Biodegradable	• Difficult to control grafting • Limited drug loading capacity – swelling issues
Poly(ε-Caprolactone) Graft Copolymer	• Biocompatible • Biodegradable • Controlled drug release • Good mechanical properties	• Limited hydrophilicity • Limited drug loading capacity • Difficult to control grafting
Dextran-β-Cyclodextrin/Ferrocene-Poly(Lactide) (Dex-β-CD/Fc-PLA)	• Biocompatible • Dual-responsive (pH and redox) • Controlled drug release • Enhanced solubility	• Complex synthesis process • Possible cytotoxicity due to ferrocene • Limited drug loading capacity
Polyurethane Graft on Linear Dextrin	• Biocompatible • Hydrophilic • Good mechanical properties • Controlled drug release	• Difficult to control grafting • Limited drug loading capacity • Potential hydrolytic degradation

payload. The micelles can release the drugs in a regulated fashion, contingent on the pH and enzymatic function within the lysosomes of cancerous cells. In particular, the copolymers are able to deliver a Bcl-2 inhibitor (ABT-199) and doxorubicin to the cancer cells for an enhanced therapeutic effect [17].

3.3.2 POLYMER-BASED DRUG DELIVERY SYSTEMS UTILIZING BOTH NATURAL AND SYNTHETIC POLYMERS

The progress of polymeric DDSs has made significant strides in recent times. These systems deliver therapeutic agents in a regulated manner and are frequently constructed from polymers that are biodegradable and bio-reducible. Alongside synthetic polymers, natural polymers, dendritic polymers, and biodegradable and bio-absorbable polymers have been researched for drug delivery applications. In addition, investigations have been conducted on macromolecular therapeutics that do not involve drugs, as well as on polymeric systems that mimic biological processes. Furthermore, both viral and non-viral vectors have been explored for their potential use in delivering genes through polymers. These carriers comprise a range of polymer types, including polyethylenimine derivatives, copolymers, and biodegradable polymers, as well as conjugates of DNA and RNA. By acting as vehicles for strong anticancer medications, biocompatible and biologically relevant copolymers, and dendrimers have also demonstrated promising outcomes in the treatment of cancer. Polymeric drug and gene delivery systems may adopt new and enhanced designs as a result of the fusion of synthetic and biological disciplines [87].

3.3.3 SELF-ASSEMBLED CONJUGATES OF CHOLINE-BASED POLY (IONIC LIQUID) GRAFTED COPOLYMERS FOR DUAL-DRUG DELIVERY

The copolymers utilized in this study comprise choline-based poly (ionic liquid) (PIL) grafted onto poly (ethylene glycol)-block-poly (lactic acid) (PEG-b-PLA). The PIL grafts have the capability to form ionic bonds with drugs such as doxorubicin and curcumin, which ultimately results in the formation of micelles with PEG-b-PLA. This unique formation enables the micelles to deliver both drugs effectively to cancer cells, leading to synergistic effects. Furthermore, the PIL grafts are pH-sensitive and can facilitate drug release in acidic environments. By evaluating graft copolymers with different quantities of IL units, researchers generated three distinct categories of carriers. Using these copolymers, they developed dual-drug systems comprising both conjugated FUS and encapsulated RIF, as well as single-drug systems containing either conjugated FUS or encapsulated RIF. By exchanging chloride anions for FUS, the copolymers could form ionic drug-carrier conjugates and effectively deliver drugs. The capacity of the copolymers to self-assemble rendered them optimal for the entrapment of the non-ionic drug RIF within polymer micelle superstructures. RIF was discovered to be favorable when combined with FUS in self-assembled conjugates to create dual-DDSs. There was no discernible difference in the RIF encapsulation efficiency between the pharmaceutical fusidate anions and the chloride anions for either conjugates or micellar carriers. The size of the predominant fraction's self-assembled particles shrank. An immediate burst release was noted during *in vitro* tests in PBS. The degree of drug release was found to fluctuate substantially (ranging from 19–81%) based on factors such as the side chain length, ionic components of the polymer, drug type, and carrier utilized. In summary, the trimethylammonium-based graft copolymers selected for this study are ideal for creating micelles or ionic conjugates to transport a single drug. They can also be utilized as cutting-edge dual-drug systems that transport two medications in separate ways (physically versus ionically) coupled via a polymer matrix. The chosen medications, RIF and FUS, have the potential to be employed in simultaneous drug co-delivery combination therapy and antibacterial treatment, including drug-resistant bacterial strains [88].

3.3.4 DRUG DELIVERY USING GRAFT COPOLYMERS BASED ON STIMULI-RESPONSIVE POLYSACCHARIDES

The copolymers utilized in this study are derived from natural polysaccharides, including cellulose, starch, chitin, chitosan, alginate, dextran, hyaluronic acid (HA), heparin, and more. These polysaccharides are then modified through grafting synthetic polymers such as PAA, poly (N-isopropyl acrylamide), PEG, and others. The resulting graft copolymers possess a range of stimuli-sensitive properties, such as temperature, pH, light, magnetic field, enzyme, and so forth, which can activate drug release from the copolymer network or matrix [89].

3.3.5 DEVELOPMENT OF SUSTAINED RELEASE BIODEGRADABLE AND NON-CYTOTOXIC CROSSLINKED CHEMICALLY HYDROGEL USING DEXTRIN AND POLY (ACRYLIC ACID)

In this investigation, a new biodegradable hydrogel with porous structure, responsive behavior, and chemical crosslinking was produced to assess its effectiveness as a long-lasting drug delivery platform for ornidazole and ciprofloxacin. pH- and temperature-responsive hydrogels utilizing crosslinked dextrin and PAA were also synthesized and analyzed for dual drug delivery. The hydrogels were created utilizing a Michael-type addition process and the N, N'-methylene bisacrylamide (MBA) crosslinker, with several reaction parameters tuned for the optimal outcome. Physicochemical techniques and rheological parameter measurements were used to evaluate the hydrogel characteristics and kinetics, while non-cytotoxicity towards human mesenchymal stem cells (hMSCs) and biodegradability were confirmed. Studies conducted outside a living organism illustrated that ornidazole and ciprofloxacin could be gradually released by the hydrogel for a period of up to 90 days. The research suggests that the dual-drug delivery system of c-Dxt/pAA hydrogels could be a viable option [90].

3.3.6 IN VITRO CONTROLLED RELEASE UTILIZING THE PH-RESPONSIVE PROPERTIES OF DEXTRAN-POLY (ACRYLIC ACID) COPOLYMER

By copolymerizing dextran and acrylic acid, Dextran-Poly (acrylic acid) copolymers (D-A) were created, and their structural characteristics and functionality were assessed using a variety of techniques. The results indicate significant interactions between the proton-acceptors of glucose units in dextran and the carboxyl groups of PAA and that the pH responsiveness of the resulting copolymers is greatly affected by the AA/dextran molar ratio. The D-A copolymer demonstrates intelligent behavior in response to pH, particularly shrinking in low-pH solutions and swelling in high-pH conditions. In the investigation of the D-A copolymer's ability to function as a drug delivery matrix, ibuprofen was employed as a model drug [91].

3.3.7 DRUG RELEASE CONTROL THROUGH NANOPARTICLE CARRIERS BASED ON POLY (ETHYLENE GLYCOL) METHACRYLATE GRAFTED ONTO CHITOSAN

This work used a Michael addition process to create a novel polymer known as chitosan grafted PEG methacrylate (CS-g-PEGMA) as a micro/nanocarrier for drug delivery. NMR and FT-IR analyses were used to validate the chemical structure of CS-g-PEGMA, and it showed enhanced solubility in an aqueous solution. CS-g-PEGMA micro/nanoparticles were created utilizing a reverse emulsion technology and a double crosslinking approach. The particles' homogeneous round shape and comparatively low polydispersity were shown by SEM examination. The swelling response of the nanoparticles to the aqueous medium was correlated to the pH-sensitive behavior, and *in vitro* tests demonstrated controlled release of bevacizumab (BEV) for several days. Hemocompatibility investigations indicated that the nanoparticles are generally safe and can be locally administered through injection. Thus, CS-g-PEGMA-based micro/nanoparticles appear to be a promising vehicle for the controlled release of BEV in ophthalmic DDSs. Lower dosages (reducing side effects), prolonged release (less frequent administration), and targeted delivery via local injection near the retina, resulting in efficient local delivery for up to 14–30 days, are the benefits of BEV-loaded micro/nanoparticles [92].

3.3.8 APPLICATION OF POLY(N-VINYLPYRROLIDONE)-BLOCK-POLY (VINYL ACETATE) AS A VEHICLE FOR DELIVERY OF HYDROPHOBIC DRUGS

The researchers created a type of polymer called poly((N-vinylpyrrolidone)-block-poly (vinyl acetate)) (PVP-b-PVAc) with different sizes and lengths to test if it could be used to carry hydrophobic riminophenazine compounds, which are effective in fighting cancer. The PVP-b-PVAc created bilayer membranes in an aqueous environment, and clofazimine was successfully encapsulated in these membranes using a dialysis technique. The researchers used transmission electron microscopy to study the drug-loaded vesicles, and they found that they were small and had a size range between 180–200 nanometers. They also tested the PVP-b-PVAc drug carrier's toxicity *in vitro* and found that it was biocompatible, meaning it didn't harm cells. These findings imply that block copolymers of PVP-b-PVAc may have utility in transporting drugs that are hydrophobic in nature [93].

3.4 RECENT ADVANCES IN GRAFT COPOLYMER-BASED DRUG DELIVERY SYSTEMS

To specifically target the release site and rate of medications in the body, DDSs combine conventional drug delivery techniques with tailored technology. This level of specificity is made possible by biodegradable and bioabsorbable polymers, which are essential components in many novel DDSs.

Because of their bioabsorbable characteristics, hydrogels comprising poly(lactic acid), poly(glycolic acid), and their copolymers are commonly employed as the delivery element in such systems. Whether the DDS employs a biodegradable implant to distribute medication deep within the body or subcutaneously, the use of biodegradable and bio-absorbable polymers provides a secure and reliable framework for delivering medication without causing harm to the body [87].

The application of polysaccharides in drug delivery has mitigated the limitations associated with the usage of artificial polymers [94]. The adoption of recombinant DNA techniques, specific microbial strains, organisms from the algal and plant realms, and sustainable sources have significantly expanded the availability of polysaccharides in diverse forms, possessing unique compositions and attributes that are challenging to replicate through laboratory synthesis. In addition to being easier to produce, these resources are less expensive than manufactured polymers. Recent scientific progress in polysaccharide hydrogel networks has led to the development of various strategies for drug delivery. Polysaccharides have shown to be incredibly adaptable, acting as polymers for controlled protein release, biocompatible scaffolds for tissue engineering, and matrices for encasing live cells. Considerable focus has been directed towards the application of polysaccharides in DDSs, which have been extensively investigated in current research endeavors. Polysaccharides have a more complex structure than other polymer classes, which enables a wider variety of chemical derivatizations and diverse uses [89].

In order to attain an ideal concentration, extend the release period, increase medication activity, and minimize adverse effects, natural polysaccharides are used as carriers for pharmaceuticals in the pharmaceutical sector [95]. When compared to synthetic polymers, polysaccharides are readily available in nature, biodegradable, nontoxic, affordable, and their structural makeup can be modified for tailored drug delivery [96].

3.4.1 MATRIX-EMBEDDED POLYSACCHARIDE FOR CONTROLLED DRUG RELEASE

Maintaining the therapeutic amount of medication in bodily tissue over the course of therapy is a difficult task in the subject of PKs. Drug release kinetics in traditional delivery systems (TDS) are instantaneous and unregulated. TDS causes a fast rise in drug concentration in various body tissues after dose administration, followed by a proportional drop. This is because the absorption of drugs in transdermal delivery systems (TDS) relies entirely on the body's ability to assimilate the therapeutic compound. As the drug concentration approaches toxic levels and then drops below the useful therapeutic level, this can result in side effects and drug resistance. Controlled release systems (CRS) strive to maintain the drug concentration within the desired tissues at an effective level for an extended duration by minimizing both excessive concentrations and subtherapeutic levels associated with TDS. Due to the decreased dosage frequency, this lessens adverse effects, raises the likelihood of medication resistance, and improves patient convenience. The most effective CRS entails encapsulating the medication within a distinctive matrix that releases it in a regulated manner, while concurrently sustaining therapeutic drug concentrations in either the bloodstream or the specific targeted tissue site [97]. Enclosing a drug in a hydrophobic or hydrophilic matrix is a common method of achieving controlled release due to its low cost and ease of fabrication. Wax, polyethylene, polypropylene, and ethyl cellulose are examples of hydrophobic matrices. Hydrophilic matrices include carboxymethyl tamarind, hydroxypropyl methylcellulose (HPMC), methylcellulose, hydroxypropyl cellulose (HPC), alginates, scleroglucan and sodium carboxymethyl cellulose (CMC) [97].

The gastrointestinal (GI) tract is composed of three sections: the stomach, small intestine, and large intestine. The GI system has a wide range of pH values, from the stomach's low pH of 1.2 to 1.4 (under fasting circumstances) to the colon's high pH of 7.4. The stomach can be a tough environment for the breakdown of peptides and other medications because of the acidic environment and many enzymes present there [98, 99]. An appropriate drug delivery matrix for oral intake should provide protection to the medication against the harsh and acidic environment of the stomach, and ideally release it within the lower GI tract. As a result, the matrix should release drugs at a reduced rate under acidic pH settings and at a larger rate under alkaline pH conditions.

A new polymeric material has been created by combining conventional and microwave-assisted techniques. The substance is made of carboxymethyl guar gum (GG) that has been grafted with polyacrylamide, and earlier studies have shown that the microwave-assisted graft copolymerization approach produces a greater percentage of grafting efficiency (%GE) than the traditional method. The mechanism of this process necessitates the employment of an initiator that generates free radical sites on the polysaccharide backbone under the influence of microwave irradiation. These free radicals facilitate the grafting of polyacrylamide onto the backbone.

To halt the process, hydroquinone is utilized as an inhibitor. Various grades of graft copolymers were produced and enhanced by modifying the reaction parameters to achieve higher %GE. The presence of graft copolymerization was validated through several characterizations [100].

3.4.2 POLYSACCHARIDES RESPONSIVE TO ELECTRICAL STIMULATION

In a recent investigation, the focus was on developing an electrically responsive transdermal DDS using a copolymer derived from a polysaccharide-based poly(acrylamide)-grafted-sodium alginate. The hydrogel exhibited a contraction and subsequent increased drug release when subjected to an electrical stimulus. The study revealed a direct correlation between the electric current density applied and the magnitude of medication release. When the electric stimulation was on and off, the copolymer also showed an on-off pulsatile drug release. According to histopathology investigations, the administered electric shock altered the stratum corneum's cellular and structural integrity, but that this alteration was reversible [101]. An electrically sensitive transdermal medication delivery system was created by the authors of a recent study employing a copolymer of polysaccharide-based poly(acrylamide) and grafted sodium alginate. When an electric stimulation was delivered, the inflated hydrogel shrank, which facilitated drug release that got better with higher electric current densities. Upon activation or deactivation of the electrical stimulus, this copolymer also demonstrated pulsatile drug release, characterized by alternating "on and off" cycles. Histopathological examinations indicated that the electrical stimulus-induced reversible alterations in the structural and cellular integrity of the stratum corneum. In a different work, the scientists created an electro-responsive transdermal delivery system for ketoprofen utilizing a hydrogel made of polysaccharide-based poly(acrylamide)-grafted xanthan gum. The drug permeation study demonstrated a notable enhancement in drug release within the electrically stimulated setting, with the intensity of the electrical current and the density of the rate-regulating membrane influencing the extent of release. Additionally, the system exhibited pulsatile drug release when the electric stimulus was activated or deactivated. Analysis of skin histology confirmed the impact of the electric stimulus on the structure of the stratum corneum and its cellular composition. In order to offer on-demand medication release that is initiated by an electric signal, these graft copolymer hydrogels might be employed efficiently [102].

3.4.3 POLYSACCHARIDES RESPONSIVE TO PH CHANGES

In recent work, Dharela et al. [103] investigated the swelling behavior of an acrylic acid-grafted hydrogel consisting of GG that was swollen using gamma radiation. The study employed a 0.9% NaCl solution and concentrated on properties of temperature, pH, and salt sensitivity. According to findings, the hydrogel swelled quickly and responded to stimuli at certain pH and temperature levels. The hydrogel demonstrated compatibility with biological systems, degradability, and economic viability, positioning it as a promising substance for applications in drug delivery and separation procedures. Polyacrylamide-g-guar gum graft copolymers, characterized by varying ratios of GG and acrylamide, were synthesized and subsequently subjected to hydrolysis to introduce carboxylic functional groups. These functionalized copolymers were utilized in the production of tablets loaded with diltiazem. The graft and hydrolyzed copolymer-infused tablets were made using wet granulation. The *in vitro* release of the drug was examined in simulated gastric and enteric conditions for both types of copolymers, and the release persisted for a duration of 8 hours in the gastric setting and 12 hours in the intestinal environment. The hydrolyzed copolymer exhibited an initial mechanism controlled by swelling in a 0.1 N hydrochloric acid solution, followed by dissolution-controlled release in pH 7.4, indicating pH-responsive behavior and potential for drug delivery at the intestinal stage. The unhydrolyzed copolymer demonstrated dissolution-controlled drug release [104].

By using a new graft copolymer of polyacrylamide-grafted-sodium alginate, scientists have created pH-sensitive particles for the regulated distribution of ketoprofen. The particles, which were created using ionotropic gelation and covalent crosslinking, displayed enhanced pH sensitivity, releasing the medication steadily in a pH 1.2 solution as opposed to a pH 7.4 buffer solution. By successfully preventing medication release in the stomach, the particles shielded the GI system from side effects such as mucosal erosion, ulcer development, and hemorrhage. This was supported by a stomach histology examination in albino rats [105]. For the transport of ketoprofen to the colon, researchers created interpenetrating network particles based on carboxymethylcellulose and grafted with polyacrylamide and sodium alginate using a similar procedure. In comparison to particles possessing both ionic and covalent bonds, those solely bonded through ionic interactions displayed erosion. The swelling of the particles and subsequent drug release experienced significant augmentation upon increasing the pH of the medium from acidic to alkaline. Under acidic conditions, drug release followed the Case II mechanism, whereas, in an alkaline dissolving medium, the release occurred through an anomalous mechanism [106]. In their investigation, the researchers explored methods to create interpenetrating network hydrogel beads with pH-sensitive properties using CMC sodium loaded with the anti-inflammatory drug ketoprofen and polyacrylamide-g-xanthan gum. Electron microscopy analysis revealed that the beads exhibited a nonporous matrix structure under acidic conditions while transitioning to a porous matrix structure under alkaline conditions. There was increased drug release and edema when the pH changed between 1.2 and 7.4. Based on a particular hypothesis, at elevated pH levels, the carboxyl functional group within the hydrogel undergoes ionization, resulting in an elevation of the osmotic pressure within the beads. Consequently, this heightened pressure prompts additional swelling and subsequent drug release [107].

The pH-sensitive microspheres for colon-specific medication administration of the anticancer medicine capecitabine were effectively created utilizing partly hydrolyzed polyacrylamide-graft-gum karaya. By employing free radical polymerization and alkaline hydrolysis techniques, the pH-responsive graft copolymer was successfully synthesized and characterized. The pH sensitivity and pulsatile swelling features of the microspheres made with pH-sensitive graft copolymer were rather noticeable. The microspheres' ability to entrap drugs varied from 77.30% to 88.74%. Based on an *in vitro* assessment of drug liberation, microspheres fabricated using the crosslinked pH-responsive graft copolymer exhibited a noteworthy deceleration in drug release within the upper GI tract and delivered a controlled and substantial drug quantity for up to 24 hours, particularly within the colon. Notably, the graft copolymer underwent swift and enhanced drug release in the presence of rat caecal content medium due to the activity of colonic bacteria. The use of glutaraldehyde as the crosslinking agent demonstrated a progressive drug release in the upper GI tract, followed by the maximal drug release in the lower GI tract, particularly at the colon level [108]. To specifically target the colon for the delivery of the anticancer drug capecitabine, pH-sensitive spray-dried microspheres composed of polyacrylamide-g-carboxymethylcellulose sodium were developed. The investigation on swelling properties revealed significant pH-responsive characteristics of the grafted copolymer. Notably, microspheres fabricated using the pH-sensitive grafted copolymer effectively achieved targeted drug release in the colon by reducing release in the upper GI tract. Conversely, microspheres composed of ungrafted carboxymethylcellulose sodium exhibited a lack of control over drug release in the small intestine and stomach. It's interesting to note that rats' caecal contents showed a quick rise in drug release, which was probably caused by colonic bacteria [109].

In a recent work, scientists created a drug delivery method by grafting graphene oxide into chitosan and polymethacrylic acid to create a matrix. The anticancer agent doxorubicin was found to be encapsulated with 78.6% competency and loaded with 93.8% capacity by the resultant graft copolymer. The investigation revealed a pH-dependent drug release pattern, characterized by a higher release rate at pH 4 and a lower release rate at pH 7.4. The efficacy of the developed formulations was evaluated using the MTT assay on MCF7 breast cancer cells, and the outcomes demonstrated that the formulations exhibited greater cytotoxicity towards cancer cells compared to the free drug. According to the study, this medication delivery method may be employed to increase treatment effectiveness while reducing doxorubicin side effects [110].

3.4.4 POLYSACCHARIDES RESPONSIVE TO CHANGES IN TEMPERATURE

A DDS for 5-fluorouracil was fabricated using a chitosan-graft-poly(N-isopropylacrylamide) copolymer that underwent self-assembly to form ionically crosslinked hollow nanospheres exhibiting thermo-sensitive behavior. The copolymer was combined with collapsed poly(N-isopropylacrylamide) to create core-shell micelles, with chitosan serving as the shell on top of LCST. The interior cavities enhanced the efficiency of drug loading by promoting interactions between the polymers and the drugs, while the size of the hollow spheres could be modified by manipulating the temperature and pH of the surrounding milieu. The drug release profile demonstrated pH sensitivity and controlled release behavior at temperatures surpassing the lower critical solution temperature (LCST). The reduction in particle size was attributed to the disruption of polymer-drug interactions. The hollow spheres disintegrated when pH was altered or ionic strength was reduced, leading to an increased drug release. Hollow nanoparticles that have been produced with environment-responsive properties have a lot of potential for application in intelligent medication delivery systems [111]. In the early 2000s, there was a growing interest in developing devices and biomaterials that could deliver bioactive compounds in response to external stimuli. One such study focused on developing temperature-sensitive beads for delivering indomethacin to the colon. They created a sodium alginate and N-isopropylacrylamide graft copolymer using microwave radiation, which they then carefully characterized. They discovered that decreasing the amount of cross-linking and graft polymer to drug lowered the amount of drug released from the beads. Furthermore, it was observed that the beads exhibited temperature responsiveness, as the drug release was more pronounced at 37°C compared to 25°C [112].

Potential smart medication delivery systems have been investigated using polymers with temperature-responsive characteristics. In a particular study, nanofibers were fabricated by electrospinning a copolymer of chitosan-graft-poly(N-isopropylacrylamide) using poly(ethylene oxide) as the substrate. The resultant hydrogel nanofibers were tested using L929 cell lines for cytotoxicity and drug release behavior at various temperatures. According to the study, the hydrogel nanofibers displayed a phase transition that was temperature-dependent, with an LCST in aqueous solutions of 32°C. The temperature of the surrounding area might be changed to alter the medication release rate. The hydrogel nanofibers exhibited favorable biocompatibility, thus showcasing their potential as smart materials for various applications, including tissue engineering and pharmaceutical transport [113]. In a recent study, researchers aimed to develop temperature-responsive nanoparticles using a copolymer of chitosan and poly(N-isopropylacrylamide) for the effective transportation of oxaliplatin to the designated tumor site. The graft copolymer was made using a dispersion copolymerization method without surfactants. According to the study, increasing the chitosan concentration led to larger drug loading percentages and quicker drug release above the LCST, while slower drug release was seen below the LCST. These findings point to the synthetic copolymer's thermo-sensitivity properties in drug delivery [114].

3.4.5 AMPHIPHILIC GRAFT POLYMER OF POLYSACCHARIDE

By altering the hydrophobic polymer chain segment within the polysaccharide molecule, it is possible to generate amphiphilic graft polymers featuring a high density of grafts and a restricted dispersion of molecular weight. The active sites on the molecular chain of the polysaccharide are commonly employed to initiate the polymerization process of polymer monomers. Controlled/living polymerization (CRP) and ring-opening polymerization (ROP) are commonly employed methods due to this attribute [115].

An alternative approach to graft hydrophobic polymers onto polysaccharides involves employing chemical processes like esterification, amidation, or thiol-Ene click chemistry. In these methodologies, the hydrophobic polymer undergoes functionalization with reactive moieties capable of engaging with the functional groups present on the polysaccharide molecule. This results in the establishment of covalent linkages between the two polymers, thereby generating an amphiphilic graft copolymer. Additional hydrophobic polymers that have been grafted onto polysaccharides encompass poly(ε-caprolactone-co-lactide) (PCLA), poly(ε-caprolactone-co-glycolide) (PCGA), and polyethylene glycol (PEG). These graft copolymers have found application in diverse fields, including drug administration, tissue regeneration, and bioanalytical devices [116]. The grafting of synthetic polymers onto polysaccharides showcased several sought-after characteristics, such as biocompatibility, biodegradability, and regulated drug liberation. The outcomes of this investigation hold significant implications for the development of efficient drug transport systems that exhibit elevated bioavailability while minimizing toxicity [117]. Another approach involves first synthesizing the hydrophobic polymer segments using an appropriate polymerization technology, such as ROP or ATRP, and then functionalizing them with energetic groups, such carboxyl or amino groups. Graft copolymers are created by connecting the functionalized hydrophobic polymers with the polysaccharide molecules using an appropriate coupling agent. By employing this approach, a more precise regulation of the arrangement and molecular weight of the hydrophobic polymer segments, along with the extent of grafting onto the polysaccharide molecules, can be achieved [118].

3.4.6 OTHER RECENT DEVELOPMENTS IN GRAFTING OF COPOLYMERS

The impact of various copolymerization conditions on copolymer properties was examined by the authors. The copolymer structure was confirmed using infrared (IR) spectroscopy. The grafting copolymerization reaction's mechanism and kinetics are thoroughly explored. The copolymer samples were analyzed for their adsorption parameters, revealing their favorable adsorptive properties for eliminating transition metals from aqueous solutions. It is still unclear, nevertheless, whether these substances are real copolymers or homopolymer mixtures made by polymerizing the monomer that adsorbs to chitosan particles [119]. To fabricate water-soluble copolymers of polyaniline (PANI), the scientists introduced an innovative approach that involved the amalgamation of grafting PSMA (poly(styrene-alt-maleic anhydride)) with PANI to yield conductive comb copolymers (PANI-g-PSMA) characterized by significantly enhanced processability. PANI nanoparticles were initially synthesized through chemical means, accompanied by ultrasonic radiation. Subsequently, they were subjected to free-radical polymerization to generate PSMA. A novel method was used to graft PANI onto the PSMA backbone, creating a conductive copolymer that resembles a comb and improves its processability [120].

An intriguing study explores the chemical oxidative polymerization of pyrrole in the presence of a macromonomer to generate polypyrrole (PPy) grafted onto a polystyrene (PS) macromonomer. The researchers implemented a meticulous purification process that involved successive dissolution of non-grafted polymer chains. The ungrafted polystyrene was initially dissolved in cyclohexane, followed by the dissolution of PPy in tetrahydrofuran (THF), resulting in the insolubility of the graft copolymer alone. This discovery exemplifies the concept of copolymerization. Notably, in comparison to PPy, the solubility and processability of the copolymer were enhanced as the PPy content increased on the functionalized PS. The synthesized graft copolymer showed adequate redox property stability, and the PPy segments retained outstanding redox activity despite being connected to a non-conjugated system [121]. By looking into the grafting of PANI onto poly(vinyl chloride), Massoumi and colleagues investigated a different variant of their methodology. In a combination containing aniline, they used the chemically produced aniline-functionalized PVC macromonomer and chemically and electrochemically polymerized both. The chemical structure of both variants of graft copolymers was verified using nuclear magnetic resonance (NMR) and infrared (IR) spectroscopy, while electrochemical and conductometric methods were employed for their evaluation. Graft copolymers showed less electrical conductivity and electroactivity than PANI, as was to be predicted [122].

NIPAM and NHMAM were successfully grafted onto alginate, resulting in the formation of a novel responsive copolymer called alginate-g-P(NIPAM-co-NHMAM). The copolymer's chemical composition was determined through analysis employing FT-IR and 1H NMR techniques. The copolymer was utilized to create a three-dimensional network-structured aerogel with thermo- and pH-sensitive characteristics. The aerogel showed excellent temperature and pH responsiveness, with up to 13.24% drug loading efficiency. Drug release was completely detected under neutral settings but was slowed down under acidic conditions. The drug release rate was elevated beyond the LCST. The investigation elucidated the drug release mechanism of the aerogel. In general, the dual stimuli-responsive aerogel exhibits promising prospects as a DDS [123]. An overview of the above-mentioned recent advances in graft copolymer-based DDSs is summarized in Table 3.3.

3.5 CHALLENGES IN GRAFT COPOLYMER-BASED DRUG DELIVERY SYSTEMS

In recent times, there has been significant interest in graft copolymer-mediated DDSs as a promising approach for regulated and site-specific administration of medications. In these systems, one or more polymer types are grafted onto a backbone polymer to create a copolymer with special characteristics that may be customized for certain drug delivery purposes. While DDSs based on graft copolymers present numerous benefits, including enhanced drug solubility, extended drug release, and increased drug effectiveness, they also encounter several hurdles that necessitate attention for their successful implementation in clinical applications.

3.5.1 SELECTION OF APPROPRIATE GRAFT COPOLYMERS

The selection of graft copolymers plays a vital role in formulating a proficient DDS. The graft copolymers should be biocompatible, biodegradable, and have suitable physicochemical properties such as molecular weight, hydrophilicity/hydrophobicity balance, and compatibility with the drug. The grafting process should also be efficient, reproducible, and scalable. However, identifying suitable graft copolymers that meet all these criteria can be challenging due to the vast array of polymers available, and the need to balance conflicting requirements, such as drug release rate, stability, and toxicity [17, 20, 89, 124, 125].

3.5.2 DRUG LOADING AND RELEASE CONTROL

Attaining an ideal drug loading capacity and achieving precise control over drug release presents a significant obstacle in graft copolymer-mediated DDSs. To achieve the desired drug release profile, the drug must be proficiently encapsulated within the graft copolymer matrix. However, several factors, such as the drug's physicochemical characteristics, the composition of the graft copolymer, and the fabrication technique employed, can influence

TABLE 3.3 Overview of Recent Advances in Graft Copolymer-Based Drug Delivery Systems

Type of Drug Delivery System	Graft Copolymer	Key Advancements
Dual-Responsive Graft Copolymer-Based Drug Delivery	Dex-β-CD/Fc-PLA	• pH and redox-responsive drug release – Enhanced solubility – Controlled drug release
Thermoresponsive Graft Copolymer for Controlled Drug Release	Starch-g-poly(N-isopropylacrylamide)	• Thermoresponsive drug release – Biodegradable – Controlled drug release
Mucoadhesive Graft Copolymer for Drug Delivery	Chitosan-g-poly (ethylene glycol)	• Mucoadhesive properties – Controlled drug release – Enhanced solubility – Biodegradable
Nanoparticles Based on Poly(ε-caprolactone) Graft Copolymer	Poly(ε-caprolactone) graft copolymer	• Biodegradable – Controlled drug release – Good mechanical properties
Polymer Brush-Based Graft Copolymer for Surface Modification	PEG graft copolymers	• Reduced immunogenicity – Reduced protein adsorption – Enhanced solubility
pH-Responsive Graft Copolymer for Cancer Drug Delivery	Dextran-Poly (acrylic acid) graft copolymer	• pH-responsive drug release – Controlled drug release – good water solubility
Self-Assembling Graft Copolymers for Drug Delivery	Poly (ethylene oxide) graft copolymers	• Self-assembly into micelles – Enhanced solubility – Reduced protein adsorption
Injectable Hydrogels Based on Polyurethane Graft on Linear Dextrin	Polyurethane graft on linear dextrin	• Injectable – Controlled drug release – Good mechanical properties – Hydrophilic

the efficiency of drug loading. Furthermore, maintaining command over the drug release tempo from the graft copolymer matrix is vital to guarantee a sustained and regulated drug administration. Ensuring accurate management of drug release kinetics, such as zero-order or first-order release patterns, can be demanding and necessitates meticulous consideration of graft copolymer characteristics, manufacturing methods, and formulation variables [126–129].

3.5.3 STABILITY AND BIODEGRADABILITY

The stability and biodegradability of graft copolymer-based DDSs are important factors that determine their safety and efficacy. The graft copolymers should remain stable during storage, transportation, and administration to maintain the integrity and functionality of the DDS. Nevertheless, certain graft copolymers might experience degradation or alterations in their physicochemical attributes with the passage of time, resulting in modifications in drug release kinetics and potential compromise of drug effectiveness. Furthermore, the biodegradability of the graft copolymer matrix plays a pivotal role in attaining regulated drug release and preventing prolonged accumulation of the polymer within the body. Achieving the right balance between stability and biodegradability is a challenge that requires careful selection of graft copolymers and formulation parameters [20, 130–133].

3.5.4 REGULATORY APPROVAL AND CLINICAL TRANSLATION

Regulatory approval and clinical translation of graft copolymer-based DDSs pose significant challenges. These systems need to undergo rigorous evaluation of their safety, efficacy, and quality to comply with regulatory requirements. The complex nature of graft copolymer-based DDSs, which often involve multiple polymers, fabrication techniques, and drug formulations, can make regulatory approval challenging. Furthermore, the transition of graft copolymer-mediated DDSs from preclinical to clinical environments can be intricate, time-intensive, and financially demanding, necessitating thorough evaluation in relevant animal models and conducting extensive clinical trials to establish their safety and effectiveness in human subjects [133–135].

3.5.5 COST-EFFECTIVENESS AND SCALABILITY

The cost-effectiveness and scalability of graft copolymer-based DDSs are crucial factors that determine their practicality for large-scale production and commercialization. The synthesis of graft copolymers and fabrication of DDSs may involve complex processes and expensive raw materials, which can add to the production costs. Achieving cost-effective and scalable production methods for graft copolymer-based DDSs while maintaining their quality, consistency, and efficacy is a significant challenge. This may require optimization of synthesis routes, formulation processes, and utilization of cost-effective raw materials without compromising the performance of the DDS [17, 89, 136].

3.5.6 TARGETING AND TISSUE-SPECIFIC DELIVERY

Achieving targeted and tissue-specific drug delivery is a key goal in drug delivery research. Graft copolymer-based DDSs offer the potential to target specific tissues or cells through surface modifications, ligand conjugation, or other targeting strategies. However, achieving effective targeting and tissue-specific delivery can be challenging due to the complexity of biological systems, variability in tissue physiology, and limitations in the availability of suitable targeting ligands or receptors. Securing the precise delivery of the drug to the desired tissue using the graft copolymer-based DDS while maintaining controlled and prolonged release necessitates meticulous design, optimization, and comprehensive *in vivo* assessment [137–140].

3.5.7 IMMUNOGENICITY AND TOXICITY CONCERNS

The immunogenicity and toxicity of graft copolymer-based DDSs are important concerns that need to be addressed for their safe clinical use. Graft copolymers or their degradation products may trigger immune responses or cause toxic effects *in vivo*, leading to adverse reactions or reduced therapeutic efficacy. Therefore, a thorough evaluation of the immunogenicity and toxicity of graft copolymer-based DDSs is crucial. This includes assessing their biocompatibility, potential for tissue irritation, systemic toxicity, and long-term effects on organs or tissues. Strategies to minimize immunogenicity and toxicity, such as using biocompatible graft copolymers, optimizing the drug release profile, and conducting comprehensive toxicity evaluations, need to be carefully considered during the design and development of graft copolymer-based DDSs [130, 141–143].

3.6 CONCLUSION

Graft copolymer-mediated DDSs offer a hopeful approach to enhance the dissolvability, durability, and precise delivery of diverse medications. These copolymers consist of a main polymer chain and appended chains of diverse polymers, allowing customization for specific purposes. Through self-assembly, graft copolymers can generate micelles, nanoparticles, or conjugates that effectively encapsulate or bind drugs and enable regulated release. Some of the advantages of graft copolymers are their biocompatibility, biodegradability, tunable amphiphilicity, and stimuli-responsiveness. While graft copolymer-based DDSs hold great promise for improving the safety and efficacy of drug delivery, they also face several challenges that need to be overcome for their successful translation into clinical practice. These challenges include selecting appropriate graft copolymers, achieving optimal drug loading and release control, ensuring stability and biodegradability, obtaining regulatory approval, addressing cost-effectiveness and scalability concerns, achieving targeted and tissue-specific delivery, managing immunogenicity and toxicity concerns, and addressing intellectual property and patenting challenges. Addressing these obstacles necessitates cross-disciplinary endeavors, encompassing pioneering investigations, rigorous analysis, comprehensive assessment, and strategic deliberation, in order to foster the creation of graft copolymer-based DDSs that are safe, efficacious, and commercially feasible for clinical implementation.

Graft copolymer-driven DDSs have unveiled fresh possibilities for the advancement of groundbreaking and efficient treatments. Nonetheless, numerous prospects exist for further enhancement and ingenuity. Expanding horizons and prospective research directions involve devising novel approaches for the synthesis and analysis of graft copolymers, enabling meticulous regulation of their molecular weight, composition, structure, and functional attributes. Another direction is to explore new types of polymers and monomers that can be used as backbone or side chains to impart desirable properties such as biocompatibility, biodegradability, stimuli-responsiveness, or targeting ability. Additionally, the efficacy and safety of graft copolymer-based DDSs should be evaluated in preclinical and clinical trials for various diseases and conditions. Ultimately, the development of intelligent and versatile DDSs based on graft copolymers should focus on the creation of adaptable and multi-purpose platforms capable of synergizing diagnostics, therapeutics, and imaging techniques. These advanced systems have the potential to revolutionize personalized medicine and cater to individualized patient needs.

KEYWORDS

- **controlled-release technologies**
- **drug delivery**
- **drug delivery system**
- **graft copolymers**
- **polyethylene glycol**
- **polyethylene oxide**

REFERENCES

1. Jain, K. K. (2020). An overview of drug delivery systems. *Drug Delivery Systems*, 1–54.
2. Ramesh, P. (1997). Transdermal delivery of drugs. *Indian Journal of Pharmacology, 29*(3), 140.
3. Rao, P. R., & Diwan, P. V. (1998). Formulation and *in vitro* evaluation of polymeric films of diltiazem hydrochloride and indomethacin for transdermal administration. *Drug Development and Industrial Pharmacy, 24*(4), 327–336.
4. Rao, P. R., & Diwan, P. V. (1997). Permeability studies of cellulose acetate free films for transdermal use: Influence of plasticizers. *Pharmaceutica Acta Helvetiae, 72*(1), 47–51.
5. Thacharodi, D., & Rao, K. P. (1995). Development and *in vitro* evaluation of chitosan-based transdermal drug delivery systems for the controlled delivery of propranolol hydrochloride. *Biomaterials, 16*(2), 145–148.
6. Krishna, R., & Pandit, J. K. (1996). Carboxymethylcellulose-sodium based transdermal drug delivery system for propranolol. *Journal of Pharmacy and Pharmacology, 48*(4), 367–370.
7. Bhat, M., Shenoy, S. D., Udupa, N., & Srinivas, C. R. (1995). Optimization of delivery of betamethasone-dipropionate from skin preparation. *Indian Drugs, 32*(5), 211–214.
8. Chien, Y. W. (1991). *Novel Drug Delivery Systems* (Vol. c. 1074). Drugs and the Pharmaceutical Sciences.
9. Langer, R. (1998). Drug delivery and targeting. *Nature, 392*(6679 Suppl), 5–10.
10. Langer, R. (1990). New methods of drug delivery. *Science, 249*(4976), 1527–1533.
11. Allen, T. M., & Cullis, P. R. (2004). Drug delivery systems: Entering the mainstream. *Science, 303*(5665), 1818–1822.
12. Gidal, B. E., DeCerce, J., Bockbrader, H. N., Gonzalez, J., Kruger, S., Pitterle, M. E., Rutecki, P., & Ramsay, R. E. (1998). Gabapentin bioavailability: Effect of dose and

frequency of administration in adult patients with epilepsy. *Epilepsy Research, 31*(2), 91–99.
13. Serajuddin, A. T. (1999). Solid dispersion of poorly water-soluble drugs: Early promises, subsequent problems, and recent breakthroughs. *Journal of Pharmaceutical Sciences, 88*(10), 1058–1066.
14. Vargason, A. M., Anselmo, A. C., & Mitragotri, S. (2021). The evolution of commercial drug delivery technologies. *Nature Biomedical Engineering, 5*(9), 951–967.
15. Sherazi, T. A. (2016). Graft polymerization. In E. Drioli & L. Giorno (Eds.), *Encyclopedia of Membranes* (pp. 1–6). Springer.
16. Bhattacharya, A., Rawlins, J. W., & Ray, P. (2008). *Polymer Grafting and Crosslinking*. John Wiley & Sons.
17. Bláhová, M., Randárová, E., Konefał, R., Nottelet, B., & Etrych, T. (2020). Graft copolymers with tunable amphiphilicity tailored for efficient dual drug delivery via encapsulation and pH-sensitive drug conjugation. *Polymer Chemistry, 11*(27), 4438–4453.
18. Rayeroux, D., Travelet, C., Lapinte, V., Borsali, R., Robin, J. J., & Bouilhac, C. (2017). Tunable amphiphilic graft copolymers bearing fatty chains and polyoxazoline: Synthesis and self-assembly behavior in solution. *Polymer Chemistry, 8*(29), 4246–4263.
19. Sakamoto, Y., & Nishimura, T. (2022). Recent advances in the self-assembly of sparsely grafted amphiphilic copolymers in aqueous solution. *Polymer Chemistry, 13*(46).
20. Chountoulesi, M., Selianitis, D., Pispas, S., & Pippa, N. (2023). Recent advances on PEO-PCL block and graft copolymers as nanocarriers for drug delivery applications. *Materials (Basel), 16*(6), 2298.
21. Yin, L., Pang, Y., Shan, L., & Gu, J. (2022). The *in vivo* pharmacokinetics of block copolymers containing polyethylene glycol used in nanocarrier drug delivery systems. *Drug Metabolism and Disposition, 50*(6), 827–836.
22. Brus, C., Petersen, H., Aigner, A., Czubayko, F., & Kissel, T. (2004). Physicochemical and biological characterization of polyethyleneimine-graft-poly (ethylene glycol) block copolymers as a delivery system for oligonucleotides and ribozymes. *Bioconjugate Chemistry, 15*(4), 677–684.
23. Sun, S., Cui, Y., Yuan, B., Dou, M., Wang, G., Xu, H., Wang, J., Yin, W., Wu, D., & Peng, C. (2023). Drug delivery systems based on polyethylene glycol hydrogels for enhanced bone regeneration. *Frontiers in Bioengineering and Biotechnology, 11*.
24. Working, P. K., et al. (1997). In *Chemistry and Biological Applications of Polyethylene Glycol (American Chemical Society Symposium Series 680)* (pp. 45–57). San Francisco.
25. Richter, A. W., & Akerblom, E. (1983). Antibodies against polyethylene glycol produced in animals by immunization with monomethoxy polyethyleneglycol modified proteins. *International Archives of Allergy and Applied Immunology, 70*(2), 124–131.
26. Cheng, T., et al. (1999). Accelerated clearance of polyethylene glycol modified proteins by anti-polyethylene glycol IgM. *Bioconjugate Chemistry, 10*(4), 520–528.
27. Kozlowski, A., & Harris, J. M. (2001). Improvements in protein PEGylation: Pegylated interferons for treatment of hepatitis C. *Journal of Controlled Release, 72*(2), 217–224.
28. Monfardini, C., et al. (1995). A branched monomethoxy polyethylene glycol for protein modification. *Bioconjugate Chemistry, 6*(1), 62–69.
29. Koslowski, A., Charles, S. A., & Harris, J. M. (2001). Development of pegylated interferons for the treatment of chronic hepatitis C. *BioDrugs, 15*(6), 419–429.

30. Katre, N. V. (1993). The conjugation of proteins with polyethylene glycol and other polymers: Altering properties of proteins to enhance their therapeutic potential. *Advanced Drug Delivery Reviews, 10*(1), 91–114.
31. Zaplinsky, S., & Lee, C. (1992). In J. M. Harris (Ed.), *Polyethylene Glycol Chemistry: Biotechnical and Biomedical Applications* (pp. 347–370). Plenum Press.
32. Işıklan, N., & Kazan, H. (2018). Thermoresponsive and biocompatible poly (vinyl alcohol)-graft-poly (N, N-diethylacrylamide) copolymer: Microwave-assisted synthesis, characterization, and swelling behavior. *Journal of Applied Polymer Science, 135*(12), 45969.
33. Rivera-Hernández, G., Antunes-Ricardo, M., Martínez-Morales, P., & Sánchez, M. L. (2021). Polyvinyl alcohol based-drug delivery systems for cancer treatment. *International Journal of Pharmaceutics, 600*, 120478.
34. Gaaz, T., Sulong, A., Akhtar, M., Kadhum, A., Mohamad, A., & Al-Amiery, A. (2015). Properties and applications of polyvinyl alcohol, halloysite nanotubes and their nanocomposites. *Molecules, 20*(12), 22833–22847.
35. Vaidya, S., Tozer, K. R., & Chen, J. (2008). An overview of embolic agents. *Seminars in Interventional Radiology, 25(3), 204–215.*
36. Córdoba-Esquivel, J. A., Guerrero-Avendaño, G. M. L., & Enríquez-García, R. (2019). Influencia de los materiales embolizantes en el tamaño de los miomas posterior a la embolización de arterias uterinas. *Revista Anales de Radiología México, 17.*
37. Marin, E., Rojas, J., & Ciro, Y. (2014). A review of polyvinyl alcohol derivatives: Promising materials for pharmaceutical and biomedical applications. *African Journal of Pharmacy and Pharmacology, 8*(10), 674–684.
38. Kobayashi, M., Toguchida, J., & Oka, M. (2003). Preliminary study of polyvinyl alcohol-hydrogel (PVA-H) artificial meniscus. *Biomaterials, 24*(4), 639–647.
39. Zhang, J., Liu, T., Liu, Z., & Wang, Q. (2019). Facile fabrication of tough photo crosslinked polyvinyl alcohol hydrogels with cellulose nanofibrils reinforcement. *Polymer, 173*, 103–109.
40. Rac, V., Lević, S., Balanč, B., Olalde Graells, B., & Bijelić, G. (2019). PVA cryogel as model hydrogel for iontophoretic transdermal drug delivery investigations: Comparison with PAA/PVA and PAA/PVP interpenetrating networks. *Colloids and Surfaces B: Biointerfaces, 180*, 441–448.
41. Pundir, C. S., Sandeep Singh, B., & Narang, J. (2010). Construction of an amperometric triglyceride biosensor using PVA membrane bound enzymes. *Clinical Biochemistry, 43*(5), 467–472.
42. Sanaeifar, N., Rabiee, M., Abdolrahim, M., Tahriri, M., Vashaee, D., & Tayebi, L. (2017). A novel electrochemical biosensor based on Fe3O4 nanoparticles-polyvinyl alcohol composite for sensitive detection of glucose. *Analytical Biochemistry, 519*, 19–26.
43. Ma'Radzi, A. H., Sugihara, S., Toida, T., & Maeda, Y. (2014). Synthesis of polyvinyl alcohol stereoblock copolymer via the combination of living cationic polymerization and RAFT/MADIX polymerization using xanthate with vinyl ether moiety. *Polymer, 55*(20), 5332–5345.
44. Segura, T., Menes-Arzate, M., León, F., Ortega, A., Burillo, G., & Peralta, R. D. (2016). Synthesis of narrow molecular weight distribution polyvinyl acetate by gamma-rays initiated RAFT/MADIX Mini emulsion polymerization. *Polymer, 102*, 183–191.

45. Göktaş, M. (2019). Synthesis and characterization of various block copolymers using PMMA-Br macroinitiator. *Chemical Papers, 73*(8), 2329–2339.
46. Koyilapu, R., Singha, S., Kutcherlapati, S. N. R., & Jana, T. (2020). Grafting of vinylimidazolium-type poly (ionic liquid) on silica nanoparticle through RAFT polymerization for constructing nanocomposite-based PEM. *Polymer, 195*, 122458.
47. Mishra, S., Rawal, A., & Nebhani, L. (2020). Imprinting the location of an in-built RAFT agent and selective grafting of polymer chains inside or outside the pores of mesoporous silica nanoparticles. *Microporous and Mesoporous Materials, 294*, 109898.
48. Chmielarz, P., Park, S., Sobkowiak, A., & Matyjaszewski, K. (2016). Synthesis of β-cyclodextrin-based star polymers via a simplified electrochemically mediated ATRP. *Polymer, 88*, 36–42.
49. Glaied, O., Dubé, M., Chabot, B., & Daneault, C. (2009). Synthesis of cationic polymer-grafted cellulose by aqueous ATRP. *Journal of Colloid and Interface Science, 333*(1), 145–151.
50. Guo, H., Li, X., Wang, J. L., Jin, X. H., & Lin, X. F. (2010). Acidic ionic liquid [NMP] H2PO4 as dual solvent-catalyst for synthesis of β-alkoxyketones by the oxa-Michael addition reactions. *Tetrahedron, 66*(36), 8300–8303.
51. Ishigaki, Y., & Mori, H. (2018). Synthesis of poly(chloroprene)-based block copolymers by RAFT-mediated emulsion polymerization. *Polymer, 140*, 198–207.
52. Tang, B. K., Li, J., Ren, Q., & Wang, C. Y. (2019). Synthesis of poly (sodium styrene sulfonate)-b-poly (butyl acrylate) block copolymers via RAFT emulsifier-free emulsion polymerization and their application in PEDOT aqueous dispersions. *Synthetic Metals, 258*, 116188.
53. Göktaş, M., & Olgun, B. (2019). One-step synthesis and characterization of poly(ε-caprolactone)-b-poly(N-isopropylacrylamide) thermo-responsive block copolymers via RAFT and ROP techniques. *Polymer Science Series B, 61*(4), 421–429.
54. Kurakula, M., & Rao, G. S. N. K. (2020). Pharmaceutical assessment of polyvinylpyrrolidone (PVP): As excipient from conventional to controlled delivery systems with a spotlight on COVID-19 inhibition. *Journal of Drug Delivery Science and Technology, 60*, 102046.
55. Peng, C., Chen, H., Wang, J., Chen, Z., Ni, M., Chen, Y., Zhang, J., & Yuan, T. (2013). Controlled degradation of polylactic acid grafting N-vinyl pyrrolidone induced by gamma ray radiation. *Journal of Applied Polymer Science, 130*(1), 704–709.
56. Franco, P., & De Marco, I. (2020). The use of poly (N-vinyl pyrrolidone) in the delivery of drugs: A review. *Polymers, 12*(5), 1114.
57. Folttmann, H., & Quadir, A. (2008). Polyvinylpyrrolidone (PVP)—One of the most widely used excipients in pharmaceuticals: An overview. *Drug Delivery Technology, 8*(2), 22–27.
58. Bothiraja, C., Shinde, M. B., Rajalakshmi, S., & Pawar, A. P. (2009). Evaluation of molecular pharmaceutical and in-vivo properties of spray-dried isolated andrographolide–PVP. *Journal of Pharmacy and Pharmacology, 61*(11), 1465–1472
59. Martins, R. M., Pereira, S. V., Siqueira, S., Salomão, W. F., & Freitas, L. A. P. (2013). Curcuminoid content and antioxidant activity in spray-dried microparticles containing turmeric extract. *Food Research International, 50*(2), 657–663.
60. Rasekh, M., Karavasili, C., Soong, Y. L., Bouropoulos, N., Morris, M., Armitage, D., Li, X., Fatouros, D. G., & Ahmad, Z. (2014). Electrospun PVP–indomethacin

constituents for transdermal dressings and drug delivery devices. *International Journal of Pharmaceutics, 473*(1–2), 95–104.
61. Fogaça, R., & Catalani, L. H. (2013). PVP hydrogel membranes produced by electrospinning for protein release devices. *Soft Materials, 11*(1), 61–68.
62. Del Consuelo, I. D., Falson, F., Guy, R. H., & Jacques, Y. (2007). *Ex vivo* evaluation of bio adhesive films for buccal delivery of fentanyl. *Journal of Controlled Release, 122*(1), 135–140.
63. Zilinskas, G. J., Soleimani, A., & Gillies, E. R. (2012). Poly (ester amide)-poly (ethylene oxide) graft copolymers: Towards micellar drug delivery vehicles. *International Journal of Polymer Science, 2012*, 1–11.
64. Shah, K. R., Chaudhary, S. A., & Mehta, T. A. (2014). Polyox (polyethylene oxide) multifunctional polymer in novel drug delivery system. *International Journal of Pharmaceutical Sciences and Drug Research, 6*(2), 95–101.
65. Kojima, H., Yoshihara, K., & Sawada, T. (2008). Extended release of a large amount of highly water-soluble diltiazem hydrochloride by utilizing counter polymer in polyethylene oxides (PEO)/polyethylene glycol (PEG) matrix tablets. *European Journal of Pharmaceutics and Biopharmaceutics, 70*(2), 556–562.
66. Thorén, L. (1980). The dextrans—Clinical data. *Developments in Biological Standardization, 48*, 157–167.
67. Cristallini, C., Barbani, N., Giusti, P., Lazzeri, L., Cascone, M. G., & Ciardelli, G. (2001). Polymerization onto biological templates, a new way to obtain bioartificial polymeric materials. *Macromolecular Chemistry and Physics, 202*(10), 2104–2113.
68. Chauvierre, C., Labarre, D., Couvreur, P., & Vauthier, C. (2003). Radical emulsion polymerization of alkylcyanoacrylates initiated by the redox system dextran–cerium (IV) under acidic aqueous conditions. *Macromolecules, 36*(16), 6018–6027.
69. Heidrick, G. W., Pippitt, C. H. Jr, Morgan, M. A., & Thurnau, G. R. (1994). Efficacy of intraperitoneal sodium carboxymethylcellulose in preventing postoperative adhesion formation. *The Journal of Reproductive Medicine, 39*(8), 575–578.
70. Tang, M., Dou, H., & Sun, K. (2006). One-step synthesis of dextran-based stable nanoparticles assisted by self-assembly. *Polymer, 47*(2), 728–734.
71. Zhang, R., Tang, M., Bowyer, A., Eisenthal, R., & Hubble, J. (2005). A novel pH-and ionic-strength-sensitive carboxy methyl dextran hydrogel. *Biomaterials, 26*(22), 4677–4683.
72. Prabaharan, M. (2008). Chitosan derivatives as promising materials for controlled drug delivery. *Journal of Biomaterials Applications, 23*(1), 5–36.
73. Kumar, M. R., Muzzarelli, R., Muzzarelli, C., Sashiwa, H., & Domb, A. J. (2004). Chitosan chemistry and pharmaceutical perspectives. *Chemical Reviews, 104*(12), 6017–6084.
74. Lakshmanan, V. K., Snima, K. S., Bumgardner, J. D., Nair, S. V., & Jayakumar, R. (2011). Chitosan-based nanoparticles in cancer therapy. In *Chitosan for Biomaterials I* (pp. 55–91). Springer.
75. Anitha, A., Sowmya, S., Sudheesh Kumar, P. T., Deepthi, S., Chennazhi, K. P., Ehrlich, H., Tsurkan, M., & Jayakumar, R. (2014). Chitosan—A versatile semi-synthetic polymer in biomedical applications. *Progress in Polymer Science, 39*(9), 1644–1667.
76. Torchilin, V. (2011). Tumor delivery of macromolecular drugs based on the EPR effect. *Advanced Drug Delivery Reviews, 63*(3), 131–135.

77. Mua, J. P., & Jackson, D. S. (1997). Fine structure of corn amylose and amylopectin fractions with various molecular weights. *Journal of Agricultural and Food Chemistry, 45*(10), 3840–3847.
78. Buleon, A., Colonna, P., Planchot, V., & Ball, S. (1998). Starch granules: Structure and biosynthesis. *International Journal of Biological Macromolecules, 23*(2), 85–112.
79. Biliaderis, C. G. (1998). Structures and phase transitions of starch polymers. In *Food Science and Technology-New York-Marcel Dekker* (pp. 57–168). Marcel Dekker.
80. Morrison, W. R., & Karkalas, J. (1990). Starch. In *Methods in Plant Biochemistry* (Vol. 2, pp. 323–352). Academic Press.
81. Ali, A. E., & AlArifi, A. (2009). Characterization and *in vitro* evaluation of starch-based hydrogels as carriers for colon specific drug delivery systems. *Carbohydrate Polymers, 78*(4), 725–730.
82. Chen, L., Li, X., Li, L., & Guo, S. (2007). Acetylated starch-based biodegradable materials with potential biomedical applications as drug delivery systems. *Current Applied Physics, 7*(4), e90–e93.
83. Kreuter, J. (1991). Nanoparticle-based drug delivery systems. *Journal of Controlled Release, 16*(1–2), 169–176.
84. Kim, H. Y., Park, S. S., & Lim, S. T. (2015). Preparation, characterization and utilization of starch nanoparticles. *Colloids and Surfaces B: Bio Interfaces, 126*, 607–620.
85. Dash, T. K., & Konkimalla, V. B. (2012). Polymeric modification and its implication in drug delivery: Poly-ε-caprolactone (PCL) as a model polymer. *Molecular Pharmaceutics, 9*(9), 2365–2379.
86. Xu, L., Wang, H., Tian, H., Zhang, M., He, J., & Ni, P. (2021). Facile construction of noncovalent graft copolymers with triple stimuli-responsiveness for triggered drug delivery. *Polymer Chemistry, 12*(14), 2152–2164.
87. Shukla, A., Maity, S., Ray, B., & Maiti, P. (2021). Dextrin and polyurethane graft copolymers as drug carriers: Synthesis, characterization, drug release, biocompatibility and in-vitro toxicity. *Carbohydrate Polymer Technologies and Applications, 2*, 100171.
88. Sung, Y. K., & Kim, S. W. (2020). Recent advances in polymeric drug delivery systems. *Biomaterials Research, 24*(1), 1, 2.
89. Niesyto, K., Mazur, A., & Neugebauer, D. (2022). Dual-drug delivery via the self-assembled conjugates of choline-functionalized graft copolymers. *Materials, 15*(12), 4457.
90. Kulkarni, R. V., Inamdar, S. Z., Das, K. K., & Biradar, M. S. (2019). Polysaccharide-based stimuli-sensitive graft copolymers for drug delivery. In *Polysaccharide Carriers for Drug Delivery* (pp. 155–177). Woodhead Publishing.
91. Das, D., Ghosh, P., Dhara, S., Panda, A. B., & Pal, S. (2015). Dextrin and poly (acrylic acid)-based biodegradable, non-cytotoxic, chemically cross-linked hydrogel for sustained release of ornidazole and ciprofloxacin. *ACS Applied Materials & Interfaces, 7*(8), 4791–4803.
92. Guo, Y., Sun, J., Bai, S., & Wu, X. (2017). pH-sensitive performance of dextran–poly (acrylic acid) copolymer and its application in controlled *in vitro* release of ibuprofen. *International Journal of Polymeric Materials and Polymeric Biomaterials, 66*(17), 900–906.
93. Savin, C. L., Popa, M., Delaite, C., Costuleanu, M., Costin, D., & Peptu, C. A. (2019). Chitosan grafted-poly (ethylene glycol) methacrylate nanoparticles as carrier for controlled release of bevacizumab. *Materials Science and Engineering: C, 98*, 843–860.

94. Bailly, N., Thomas, M., & Klumperman, B. (2012). Poly (N-vinylpyrrolidone)-block-poly (vinyl acetate) as a drug delivery vehicle for hydrophobic drugs. *Biomacromolecules, 13*(12), 4109–4117.
95. Deshmukh, A. S., Setty, C. M., Badiger, A. M., & Muralikrishna, K. S. (2012). Gum ghatti: A promising polysaccharide for pharmaceutical applications. *Carbohydrate Polymers, 87*(2), 980–986.
96. Parmar, N. S., & Shivprakash, B. (1997). Biopharmaceutical and pharmacokinetic considerations. In J. K. Jain (Ed.), *Controlled and Novel Drug Delivery* (1st ed., pp. 1–2). CBS Publishers and Distributors.
97. Bhardwaj, T. R., Kanwar, M., Lal, R., & Gupta, A. (2000). Natural gums and modified natural gums as sustained-release carriers. *Drug Development and Industrial Pharmacy, 26*(9), 1025–1038.
98. Pal, S., & Das, R. (2013). Polysaccharide-based graft copolymers for biomedical applications. In *Polysaccharide Based Graft Copolymers* (pp. 325–345). Springer.
99. Sen, G., & Pal, S. (2009). A novel polymeric biomaterial based on carboxymethyl starch and its application in controlled drug release. *Journal of Applied Polymer Science, 114*(5), 2798–2805.
100. Friend, D. R. (1991). Colon-specific drug delivery. *Advanced Drug Delivery Reviews, 7*(1), 149–199.
101. Pal, S., Ghorai, S., Dash, M. K., Ghosh, S., & Udayabhanu, G. (2011). Flocculation properties of polyacrylamide grafted carboxymethyl guar gum (CMG-g-PAM) synthesized by conventional and microwave-assisted methods. *Journal of Hazardous Materials, 192*(3), 1580–1588.
102. Kulkarni, R. V., Setty, C. M., & Sa, B. (2010). Polyacrylamide-g-alginate-based electrically responsive hydrogel for drug delivery application: Synthesis, characterization, and formulation development. *Journal of Applied Polymer Science, 115*(3), 1180–1188.
103. Kulkarni, R. V., & Sa, B. (2009). Polyacrylamide-grafted-alginate-based pH-sensitive hydrogel beads for delivery of ketoprofen to the intestine: *In vitro* and *in vivo* evaluation. *Journal of Biomaterials Science, Polymer Edition, 20*(2), 235–251.
104. Dharela, R., Raj, L., & Chauhan, G. S. (2012). Synthesis, characterization, and swelling studies of guar gum-based pH, temperature and salt responsive hydrogels. *Journal of Applied Polymer Science, 126*(1), E260–E265.
105. Toti, U. S., & Aminabhavi, T. M. (2004). Modified guar gum matrix tablet for controlled release of diltiazem hydrochloride. *Journal of Controlled Release, 95*(3), 567–577.
106. Kulkarni, R. V., & Sa, B. (2009). Electro responsive polyacrylamide-grafted-xanthan hydrogels for drug delivery. *Journal of Bioactive and Compatible Polymers, 24*(4), 368–384.
107. Kulkarni, R. V., & Sa, B. (2008). Evaluation of pH-sensitivity and drug release characteristics of (polyacrylamide-grafted-xanthan)-carboxymethyl cellulose-based pH-sensitive interpenetrating network hydrogel beads. *Drug Development and Industrial Pharmacy, 34*(12), 1406–1414.
108. Kulkarni, R. V., & Sa, B. (2008). Novel pH-sensitive interpenetrating network hydrogel beads of carboxymethylcellulose-(polyacrylamide-grafted-alginate) for controlled release of ketoprofen: Preparation and characterization. *Current Drug Delivery, 5*(3), 256–264.
109. Alange, V. V., Birajdar, R. P., & Kulkarni, R. V. (2017). Functionally modified polyacrylamide-graft-gum karaya pH-sensitive spray dried microspheres for colon

targeting of an anti-cancer drug. *International Journal of Biological Macromolecules, 102*, 829–839.
110. Alange, V. V., Birajdar, R. P., & Kulkarni, R. V. (2019). Novel spray dried pH-sensitive polyacrylamide-grafted-carboxymethylcellulose sodium copolymer microspheres for colon targeted delivery of an anti-cancer drug. *Journal of Biomaterials Science, Polymer Edition, 28*(2), 139–161.
111. Abbasian, M., Rodi, M. M., Mahmoodzadeh, F., Eskandani, M., & Jaymand, M. (2018). Chitosan-grafted-poly (methacrylic acid)/graphene oxide nanocomposite as a pH-responsive de novo cancer chemotherapy nano system. *International Journal of Biological Macromolecules, 118*, 1871–1879.
112. Li, X., Kanjwal, M. A., Lin, L., & Chronakis, I. S. (2013). Electrospun polyvinyl-alcohol nanofibers as oral fast-dissolving delivery system of caffeine and riboflavin. *Colloids and Surfaces B: Bio Interfaces, 103*, 182–188.
113. Nuran, I., & Kucukbalc, G. (2012). Microwave-induced synthesis of alginate-graft-poly(N-isopropylacrylamide) and drug release properties of dual pH-and temperature-responsive beads. *European Journal of Pharmaceutics and Biopharmaceutics, 82*(2), 316–331.
114. Yuan, H., Li, B., Liang, K., Lou, X., & Zhang, Y. (2014). Regulating drug release from pH-and temperature-responsive electrospun CTS-g-PNIPAAm/poly (ethylene oxide) hydrogel nanofibers. *Biomedical Materials, 9*(5), 055001.
115. Patil, A. S., & Gadad, A. P. (2018). Development and evaluation of high oxaliplatin loaded CS-g-PNIPAAm co-polymeric nanoparticles for thermo and pH-responsive delivery. *Indian Journal of Pharmaceutical Education and Research, 52*(1), 62–70.
116. Atanase, L. I., Desbrieres, J., & Riess, G. (2017). Micellization of synthetic and polysaccharides-based graft copolymers in aqueous media. *Progress in Polymer Science, 73*, 32–60.
117. Wen, Y., & Oh, J. K. (2015). Recent strategies to develop polysaccharide-based nanomaterials for biomedical applications. *Macromolecular Rapid Communications, 35*(22), 1819–1832.
118. Hu, Y., Li, Y., & Xu, F. J. (2017). Versatile functionalization of polysaccharides via polymer grafts: From design to biomedical applications. *Accounts of Chemical Research, 50*(2), 281–292.
119. Wang, Q. Y., Su, H., Xia, C. C., Sun, J. Y., Liu, C., Wang, Z. Y., Gong, Q. Y., Song, B., Gao, F. B., Ai, H., et al. (2009). Amphiphilic dextran/magnetite nanocomposites as magnetic resonance imaging probes. *Chinese Science Bulletin, 54*(15), 2925–2933.
120. Abd El-Salam, H. M., Kamal, E. H. M., & Ibrahim, M. S. (2017). Synthesis and characterization of chitosan-grafted-poly(2-hydroxyaniline) microstructures for water decontamination. *Journal of Polymer Environment, 25*(4), 973–982.
121. Heydari, M., Moghadam, P. N., Fareghi, A. R., Bahram, M., & Movagharnezhad, N. (2015). Synthesis of water-soluble conductive copolymer based on polyaniline. *Polymer Advanced Technologies, 26*(2), 250–254.
122. Hatamzadeh, M., Mohammad-Rezaei, R., & Jaymand, M. (2015). Chemical and electrochemical grafting of polypyrrole onto thiophene-functionalized polystyrene macromonomer. *Materials Science in Semiconductor Processing, 31*, 463–470.
123. Massoumi, B., Mohammad-Rezaei, R., & Jaymand, M. (2016). Chemical and electrochemical grafting of polyaniline onto poly (vinyl chloride): Synthesis,

characterization, and materials properties. *Polymer Advanced Technologies, 27*(9), 1056–1063.
124. Shao, L., Cao, Y., Li, Z., Hu, W., Li, S., & Lu, L. (2018). Dual responsive aerogel made from thermo/pH sensitive graft copolymer alginate-g-P(NIPAM-co-NHMAM) for drug-controlled release. *International Journal of Biological Macromolecules, 114*, 1338–1344.
125. Maiti, S., Ranjit, S., & Sa, B. (2010). Polysaccharide-based graft copolymers in controlled drug delivery. *International Journal of PharmTech Research, 2*(2), 1350–1358.
126. Niesyto, K., & Neugebauer, D. (2020). Synthesis and characterization of ionic graft copolymers: Introduction and *in vitro* release of antibacterial drug by anion exchange. *Polymers, 12*(9), 2159.
127. Wang, D., Tan, J., Kang, H., Ma, L., Jin, X., Liu, R., & Huang, Y. (2011). Synthesis, self-assembly and drug release behaviors of pH-responsive copolymers ethyl cellulose-graft-PDEAEMA through ATRP. *Carbohydrate Polymers, 84*(1), 195–202.
128. Dinari, A., Abdollahi, M., & Sadeghizadeh, M. (2021). Design and fabrication of dual responsive lignin-based nanogel via "grafting from" atom transfer radical polymerization for curcumin loading and release. *Scientific Reports, 11*(1), 1962.
129. Muschert, S., Siepmann, F., Leclercq, B., Carlin, B., & Siepmann, J. (2009). Drug release mechanisms from ethylcellulose: PVA-PEG graft copolymer-coated pellets. *European Journal of Pharmaceutics and Biopharmaceutics, 72*(1), 130–137.
130. Du, H., Han, R., Tang, E., Zhou, J., Liu, S., Guo, X., & Wang, R. (2018). Synthesis of pH-responsive cellulose-g-P4VP by atom transfer radical polymerization in ionic liquid, loading, and controlled release of aspirin. *Journal of Polymer Research, 25*(9), 1–9.
131. Najer, A., Wu, D., Nussbaumer, M. G., Schwertz, G., Schwab, A., Witschel, M. C., Schäfer, A., Diederich, F., Rottmann, M., Palivan, C. G., & Beck, H. P. (2016). An amphiphilic graft copolymer-based nanoparticle platform for reduction-responsive anticancer and antimalarial drug delivery. *Nanoscale, 8*(31), 14858–14869.
132. Kumar, D., Gihar, S., Shrivash, M. K., Kumar, P., & Kundu, P. P. (2020). A review on the synthesis of graft copolymers of chitosan and their potential applications. *International Journal of Biological Macromolecules, 163*, 2097–2112.
133. Quynh, T. M., Yoneyama, M., Maki, Y., & Dobashi, T. (2012). Poly (N-isopropylacrylamide-co-hydroxyethyl methacrylate) graft copolymers and their application as carriers for drug delivery system. *Journal of Applied Polymer Science, 123*(4), 2368–2376.
134. Li, J., Yang, X. L., Liu, Y. H., Wu, W. X., Liu, B. Y., Wang, N., & Yu, X. Q. (2018). Chemoenzymatic synthesis of dual-responsive graft copolymers for drug delivery: Long-term stability, high loading and cell selectivity. *Journal of Materials Chemistry B, 6*(43), 6993–7003.
135. Li, W., Tang, J., Lee, D., et al. (2022). Clinical translation of long-acting drug delivery formulations. *Nature Reviews Materials, 7*(5), 406–420.
136. Hawthorne, D., Pannala, A., Sandeman, S., & Lloyd, A. (2022). Sustained and targeted delivery of hydrophilic drug compounds: A review of existing and novel technologies from bench to bedside. *Journal of Drug Delivery Science and Technology, 78*, 103936.
137. Bhosale, R. R., Osmani, R. A., Lila, A. S., Khafagy, E. S., Arab, H. H., Gowda, D. V., Rahamathulla, M., Hani, U., Adnan, M., & Gangadharappa, H. V. (2021). Ghatti gum-base graft copolymer: A plausible platform for pH-controlled delivery of antidiabetic drugs. *RSC Advances, 11*(24), 14871–14882.

138. Alsaggar, M., & Liu, D. (2018). Organ-based drug delivery. *Journal of Drug Targeting, 26*(5–6), 385–397.
139. Patra, J. K., Das, G., Fraceto, L. F., et al. (2018). Nano based drug delivery systems: Recent developments and future prospects. *Journal of Nanobiotechnology, 16*(1), 71.
140. Zhao, Z., Ukidve, A., Kim, J., & Mitragotri, S. (2020). Targeting strategies for tissue-specific drug delivery. *Cell, 181*(1), 151–167.
141. Patel, S., Bhadoria, A., & Patani, P. (2022). Colon targeted drug delivery systems: Based on polymers. *Journal of Pharmaceutical Negative Results, 13*(1), 1980–1988.
142. Shukla, A., Singh, A. P., & Maiti, P. (2021). Injectable hydrogels of newly designed brush biopolymers as sustained drug-delivery vehicle for melanoma treatment. *Signal Transduction and Targeted Therapy, 6*(1), 63.
143. Bielas, R., Mielańczyk, A., Skonieczna, M., Mielańczyk, Ł., & Neugebauer, D. (2019). Choline supported poly (ionic liquid) graft copolymers as novel delivery systems of anionic pharmaceuticals for anti-inflammatory and anti-coagulant therapy. *Scientific Reports, 9*(1), 14410.
144. Mustafai, A., Zubair, M., Hussain, A., & Ullah, A. (2023). Recent progress in proteins-based micelles as drug delivery carriers. *Polymers, 15*(4), 836.
145. Wei, X., Gong, C., Gou, M., Fu, S., Guo, Q., Shi, S., Luo, F., Guo, G., Qiu, L., & Qian, Z. (2009). Biodegradable poly(ε-caprolactone)–poly (ethylene glycol) copolymers as drug delivery system. *International Journal of Pharmaceutics, 381*(1, 2), 1–8.

CHAPTER 4

Latest Breakthrough and Obstacles in Drug Delivery Systems Utilizing Graft Copolymers

JOHNSON O. OLADELE,[1,2] BANKOLE E. OFENIFORO,[3] ISAAC O. BABATUNDE,[4] and OLUWASEUN T. OLADELE[4]

[1]*Biochemistry Unit, Department of Chemical Sciences, Faculty of Sciences, Kings University, Ode-Omu, Osun, Nigeria*

[2]*Phytochemistry and Phytomedicine Research Unit, Royal Scientific Research Institute, Osun, Nigeria*

[3]*Department of Biochemistry, Faculty of Life Sciences, University of Ilorin, Ilorin, Kwara, Nigeria*

[4]*Phytomedicine and Molecular Toxicology Research Laboratories, Department of Biochemistry, Osun State University, Osogbo, Nigeria*

ABSTRACT

The process or method of providing a pharmaceutical substance to have a therapeutic effect on individuals or animals is known as drug delivery. A fast-developing topic in medicine, pharmacology, and other related disciplines is the discovery/creation of medication delivery systems based on both natural and synthetic polymers. Polymers have long been useful excipients for traditional dosage forms. They have also performed admirably in the parenteral setting, and they are now able to conduct complex and advanced tasks, including controlled drug release and drug targeting. Numerous innovative medication delivery methods have been created as a result of developments in polymer science. Polymers demonstrate certainty in meeting the

Graft Copolymers of Polysaccharides for Drug Delivery Systems.
Rishabha Malviya and Sonali Sundram (Eds.)

demanding criteria set for these delivery systems, including the capacity to be delivered into an aerosol, biocompatibility, stability against forces generated during aerosolization, targeting of specific sites, release of the drug in a predetermined manner, and degradation within a reasonable amount of time.

4.1 INTRODUCTION

One of the cutting-edge research fields is the hunt for innovative drug delivery system (IDDS) techniques and new mechanisms of action, as these offer substantial increases in therapeutic effectiveness and bioavailability based on multidisciplinary scientific approaches at the sites of drug administration [1, 2]. A DDS integrates engineered technology with one or more conventional medication administration methods. The systems enable the capacity to target a drug's release from the body at a particular location and/or at a particular rate.

Many innovative medication delivery techniques are made possible by biodegradable and bioabsorbable polymers. The delivery composition of the schemes has been developed using bioabsorbable polymers, such as hydrogels made of poly(lactic acid) and poly(glycolic acid), and their copolymers [3, 4]. Whether the DDS relies on a biodegradable implant to distribute medication subcutaneously or deeply within the body, the biodegradable and bioabsorbable polymers provide a safe base for drug administration without harming the body.

A formulation or a tool that makes it possible to introduce a therapeutic substance into the body is known as a DDS. By modifying the pace, window of time, and location of medication release throughout the body, it increases safety and effectiveness. Medication delivery has advanced significantly over the past 20 years, but controlling medication access into some parts of the body, including the brain, is still a challenge. However, recent developments in the study of nano-DDSs' carrier-mediated passage across the blood-brain barrier are starting to offer a rationale for managing drug distribution to the brain. The uptake transporters for natural nutrients such as amino acids, peptides, hexose, monocarboxylate, and stem cells are the transport mechanisms at the blood-brain barrier [5–7]. The polymeric drugs and gene delivery methods of both synthetic and natural polymers to construct pharmaceuticals into the backbone structures in diverse instances have been examined in this current chapter. Future research potential for useful applications has also been put forth for the advancement of the fields.

4.2 NOVEL DRUG DELIVERY SYSTEMS

Innovative systems for drug delivery give a more logical approach to the creation of an ideal DDS since they are a new system that increases and broadens our knowledge of the PK and pharmacodynamic behavior of the drug. For a longer period of time, the IDDS are carriers that keep the drug concentration in therapeutic limits. Novel DDSs have a number of benefits over traditional medication delivery, including the following:

1. Over a lengthy period of time, the circulatory system or a tissue's optimal therapeutic medication concentration may be maintained.
2. A drug's predetermined rate, which aids in extending its action.
3. Drugs with a brief half-life might be increased.
4. The risk of adverse effects may be reduced by focusing on the location of action.
5. Less frequent dosing and medication waste could be achieved.
6. Increased patient adherence.

Many DDS have been developed to decreased drug loss, prevent adverse drug reaction, increase drug bioavailability, support and promote drug accumulation in required bio-zone (site), and some more are in the pipeline. Many new carriers have been discovered and demonstrated to be effective in controlled and long-term medication delivery. It is important to analyze the various terms used to describe the various main categories of novel

- Regulated or sustained DDSs provide long-term or constant (zero-order) release correspondingly at therapeutically effective blood levels to deliver drug activity at a predetermined rate.
- Localized DDSs provide drugs by regulating their rate of release close to the target.
- Control the molecular diffusion of drug molecules in the systemic circulation by designing a system that changes the release of drug molecules at a predetermined rate, providing pharmacological activity.
- Targeting the DDS is to deliver drugs through carriers for either passive diffusion or active diffusion, one base, or self-programmed approaches. These approaches are typically employed in conjunction with proper sensory equipment that can identify the receptor at the targeted region.

4.2.1 POLYMERS IN DRUG DELIVERY SYSTEMS

Polymer materials and its product global marketing are expanding quickly, and polymer products are currently one of the chemical products with the biggest volume produced worldwide. Whether in capsule or tablet formulations, polymers have always been a valuable excipient [8]. Research evidence on its usage has shown remarkable results in parenteral settings as blood circulation time modulators [9] and is now capable of providing advanced and sophisticated functions such as controlled drug delivery and drug targeting [10]. The discovery of blending polymer mixtures has been facilitated in recent decades by an ever-increasing desire for enhanced polymer characteristics [11, 12].

A new class of polymers has surfaced, and this is based on the mixing of natural and synthetic polymers, alone or in combination, to increase mechanical strength and overcome low biological performance. A combination of two or more polymers within a network, which is termed an Interpenetrating polymers network (IPN), has evolved; at least one of the polymers in the network was created while the second was being synthesized [13]. In this, the polymeric chains from the other system become entangled or penetrate the network formed by the first polymer in the network, this results in physically crosslinked networks. Each network retains its own unique properties, allowing synergistic increases in properties such as toughness or strength [14].

As a matter of fact, this IPN can swell but not dissolve in solvents, and it can be inhibited during creep, and the flow allows it to be distinguished from a polymer blend [15]. Additionally, they differ from polymer complexes and graft copolymers that contain either chemical linkages or modest levels of crosslinking. IPN can be broadly referred to as "polymer alloys" from this perspective alone, allowing polymer mixtures to be chemically compatible and thus provide the appropriate phase morphology [16]. IPN differs from other multiple systems in that it has a bicontinuous structure that, depending on the composition and degree of crosslinking, produces a material with improved properties. This structure is ideally created by the interlinking of two polymers that are in close proximity and not in contact (Figure 4.1).

The first interpenetrating polymer network (IPN) was created by Aylsworth in 1914, which is at least when the idea of an IPN first emerged [17]. This was a blend of phenol-formaldehyde resins, natural rubber, and sulfur. In a 1960s scientific investigation on polystyrene networks, Miller coined

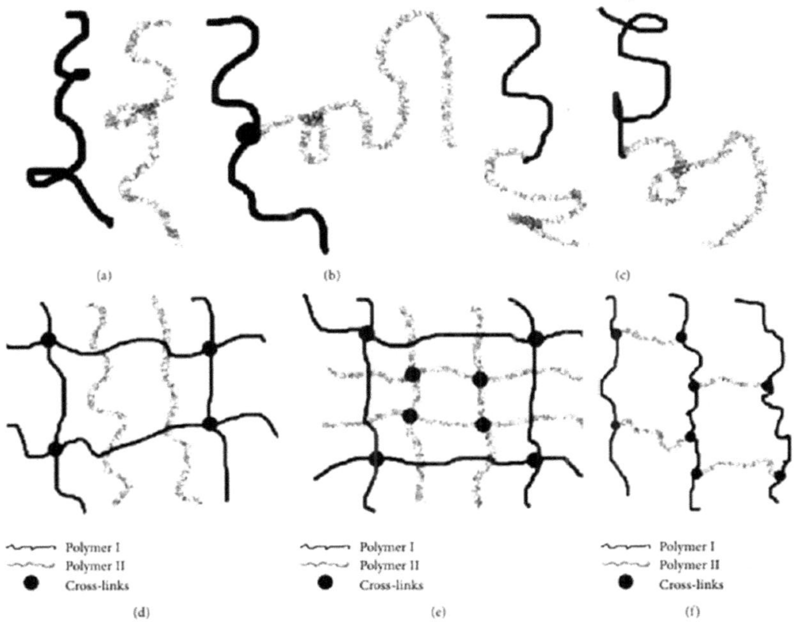

FIGURE 4.1 (a) Showing a polymer blend, (b) showing a graft copolymer, (c) showing a block copolymer, (d) showing semi-IPN, (e) showing full IPN, and (f) crosslinked copolymer.

the term "IPN" for the first time [18]. Since then, the field of IPN has grown significantly. Numerous innovative medication delivery systems have been created as a result of developments in polymer science. IPNs have outperformed traditional individual polymers, and as a result, the applications for this class of materials have expanded quickly. The pharmaceutical industry has paid close attention to IPNs' superior features, particularly in the field of medication delivery. These bio-compatible, non-toxic, and bio-degradable polymer networks are playing an increasingly important role in the delivery of bioactive compounds, particularly in regulated and focused drug delivery applications. Numerous studies have demonstrated that IPN-based delivery devices can efficiently distribute a range of medications (Table 4.1).

4.2.2 BIOREDUCIBLE POLYMERS WITH POLYETHYLENIMINE CONJUGATION

To produced bioreducible polymers with conjugated polyethylenimine, the reagent used is Traut's reagents, as this synthesized the products between PEI 1.8 kDa and poly (cystamine-bis-(acrylamide) diaminohexane) [poly

TABLE 4.1 Delivery of Variety of Medications Through Various IPN Based Novel Carriers

Sl. No.	Medication	Pharmacological Significance	Carrier System	Inference
1	Ciprofloxacin hydrochloride	Antibacterial	IPN scaffold	For quick and regulated wound healing
2	Prazosin hydrochloride	Antihypertensive	Hydrogel	Extended drug release through transdermal route
3	Metformin hydrochloride	Hypoglycemic	Mucoadhesive microspheres	Prolonged gastric residence with sustained release
4	Cefadroxil	Antimicrobial	Microgels	pH-dependent extended drug release
5	Insulin	Hypoglycemic	Microparticles	Innovative carrier for oral insulin delivery
6	Bovine serum albumin	Serum albumin	Hydrogel beads	Improved drug release in physiological saline solution
7	Cloxacillin	Antibiotic	Floating beads	Prolonged sustained release in simulated gastric fluid
8	Clarithromycin	Anti-*Helicobacter pylori*	Hydrogels	Stomach-specific drug delivery
9	Simvastatin	Hypolipidemic	IPN beads	Controlled drug release
10	Ofloxacin hydrochloride	Antibacterial	Beads	Sustained drug release
11	Chlorpheniramine maleate	Antihistamines	Beads	Controlled drug delivery
12	Aspirin	Anti-inflammatory	Microparticles	Intestinal-specific drug delivery
13	Diclofenac sodium	Anti-inflammatory	Microbead	Intestine-specific drug release with minimum gastric side effects of drug
14	5-Fluorouracil	Antimetabolite	Hydrogels	pH sensitivity to localize drug delivery
15	Buflomedil hydrochloride		Microspheres	Controlled oral delivery of highly water-soluble drug
16	5-Fluorouracil and diclofenac sodium	Antimetabolite and anti-inflammatory	Hydrogels	Localized drug delivery

(CBA-DAH)] in order to form the disulfide link [19]. Poly(CBA-DHA)-PEI can be used as a carrier gene for gene transfer. Because of steric constraints due to the P-PEI conjugation, poly(CBA-HAH), and poly(PEI), proton peaks were shifted downfield for product identification. Additionally, the ratio of the integration of the proton spectra peaks in CH_2 of PEI and poly (CBA-DAH) ($NCH_2CH_2CH_2CH_2$-$CH_2CH_2NH_2$) has been used to calculate the conjugation ratio of PEI to the PCDP.

PCDP polyplexes exhibit a 10 times greater gene transfection efficacy than Lipofectamine® polyplexes. Finally, it is determined that the bio-reducible PEI (1.8 kDa) conjugated poly (CBA-DAH) is an effective polymeric gene delivery carrier [20, 21]. The PEI (1.8 kDa)-PCDP produced in our lab has been determined to be one of the top options for mRNA, siRNA, and pDNA carriers for effective gene delivery systems [22].

Structurally, the low molecular weight (LMW) PEI was joined through disulfide bonds to poly (CBA-DAH); this PEI-poly [PCDP] conjugate interacts with pDNA and forms nano-scale polyplexes with a positive surface charge in bio-mimetic in vivo conditions. PCDP polyplexes exhibit 10-fold higher gene transfection efficiency compared to Lipofectamine-conjugated polyplexes. The report finalized that the PEI-poly [PCDP] conjugate is potent as a carrier for delivering polymeric genes [20, 21]. An analytical report from Rai et al. [22] shows that the bio-reducible PEI-poly [PCDP] conjugate synthesized in their laboratory is one of the top options for polymeric gene delivery carriers for mRNA, siRNA, and pDNA [22].

Over the past 10 years, some synthetic bio-reducible polymers have evolved and developed for gene therapy, most notably among them of great significance are poly(dimethyl-aminoethyl methacrylate), poly(amidoamine) dendrimers, diethyl-aminoethyldextran, poly(L-lysine), linear and branched polyethyleneimine (PEI), and poly(L-ornithine) [23].

4.2.3 POLYETHYLENIMINE COPOLYMERS

One method of reducing PEI's cytotoxicity is the addition of poly (ethylene glycol) (PEG) blocks to PEI. However, it is widely known that PEGylated PEI has worse transfection effectiveness than the equivalent PEI to some extent. We synthesized and tested new ABA triblock copolymers as a polymeric transfectant, using the PEG tag as the B block and linear PEI tagged ass the A block. Comparing the PEI-PEG-PEI triblock copolymers to high molecular weight PEIs, they also showed an enhanced safety profile. An intriguing new family of non-viral gene delivery techniques is the linear PEI-PEG-PEI triblock copolymers [23]. The perfect gene carrier was created using polyethyleneimine-alt-poly (ethylene glycol) copolymers for both safety and transfection effectiveness. Plasmid DNA was complexed with the copolymers. Even at high copolymer concentrations, the resultant complexes had no harmful impact on cell viability [24].

The PEG molecular weight had an impact on its transfection competence. The transfection competence of PEI 25 K is greater in MG63 and HepG2 cells was greater than that of PEI 25 K, but less in HeLa cells [25]. This is because the LMW polyglutamic acid (PEG)-Bioplastically has been attached by Aminolysis to Biodegradable Backbone Polyglutamic Acids (PEG-B-PBLG)-Coupled PEIs to create Biodegradable Gene Vectors offering excellent transfection efficiency and low cytotoxicity [26]. PEG-g-PEI-g-poly (dimethylaminoethyl Lglutamine) (PEG-g-PEI-g-PDMAEG) is a group of tri-block copolymers that were developed and tested as innovative vectors for gene therapy [27]. This created tri-block copolymers, which are, therefore, promising delivery vehicles for non-viral genes.

4.2.4 POLYETHYLENIMINE DERIVATIVES

A family of cationic polymers known as polyethyleneimine (PEI) has already been established to be proficient in gene delivery [28]. Among them, the 25 kDa branched poly (ethylenimine) (PEI) is known to be a powerful gene transport agent with exceptional gene condensation ability and excellent endosome escape activity [29]. LMW PEI (1.8 kDa) was converted into a biologically reducible polyethylene-imine (PEI (s-s-)) for effective gene delivery. It has been predicted that the bio-reducible core molecules will lessen the cytotoxicity of the copolymers and also greatly increase the molecular weights. Comparing PEI (s-s-) polyplexes to branching PEI 25 kDa, Lipofectamine® 2000 revealed that the latter had reduced cytotoxicity and higher transfection effectiveness. Additionally, with a zeta-potential value of + 34 mV and a size of 61 nm, stable polyplexes were made using PEI (s-s-) derivatives (16 kDa) [30]. The primary barrier to the application of polyethylenimine (PEI) is its cytotoxicity. Due to its strong buffering capacity, the most common cationic polymer, polyethyleneimine (PEI), has a high transfection competence. According to reports, PEI is non-degradable and cytotoxic in a variety of cell types. Polyethyleneimine copolymers were initially used in gene delivery systems to address the issues [31].

4.2.5 BIOMIMETIC AND BIO-INSPIRED POLYMERS

When used in drug delivery, bio-inspired and biomimetic systems enhance biocompatibility. The shape, surface, structure, motion, and delivery method of such a DDS all affect its efficacy. These systems have a considerable impact on biological systems as a result of their low toxicity, high biocompatibility, high interaction, and other features. [32, 33]. The innovative dendritic polymer-based medicament delivery systems at nanoscales reported here have a great deal of potential to improve therapeutic indices for cancer therapy while having minimal adverse effects [34–36]. Even though synthetic drug carriers have many uses, it is still vital to look at natural particles, such as pathogens, and the workings of mammalian cells. Due to their distinct chemical and physical characteristics, biocompatible polymeric nanoparticles are incredibly intriguing carrier candidates for the transport of medicines and DNA [37, 38].

4.2.6 BIODEGRADABLE AND BIO-ABSORBABLE POLYMERS

When just temporary implant presence is required, the drug delivery bioabsorbable devices are a superior option for the application of drug carriers [39]. Aliphatic polyesters, including polymers of lactic acid, caprolactone, glycolic acid, and dioxanone, are the synthetic and biodegradable polymers that are most frequently employed. Several kinds of polymers, including polymer esters, biodegradable carbonates, orthoesters, and anhydrides, have also been suggested as suitable implantable pharmaceutical delivery systems [40–42]. The polymers of phosphodiester, alkyl cyanoacrylates, hyaluronic acids, anhydrides, amides, ester amides, α-hydroxy acids, and natural sugars like chitosan are among the numerous degradable polymer types that are frequently used. DDSs prefer synthetic biodegradable polymers over biodegradable polymers derived from natural polymers because they are immunogenic [43–45].

4.2.7 POLY(N-(2-HYDROXYPROPYL) METHACRYLAMIDES

RAFT stands for reversible addition-fragmentation chain transfer (RAFT), on copolymerization produces degradable multiple blocks of (PTX) and N-(2-hydroxypropyl) methacrylamide (HPMA) conjugates, which are used for preclinical study, was used [46]. Conjugates of GDC0980, cytarabine, and poly (HPMA) copolymer were created. Studies conducted *in vitro* revealed that both conjugates were highly toxic to cells and that they worked well together, offering a potential chemotherapeutic approach [47]. Raft polymerization of RAFT was made possible by the use of a new bifunctionally transferable chain transfer agent (CTA) consisting of an enzyme-degradable oligo-peptide sequence to be used in the conjugation of telechemically solubilized water soluble HPMA co-polymer and HPMA co-doxyribicin conjugates (DOX) [48].

4.2.8 DENDRITIC POLYMERS

Dendritic polymers, in nature, are viscous, low solution, strongly soluble, have a huge population of terminal functional groups, and highly branched polymers with controlled topologies [49]. Through the use of synthetic processes and the degree of branching, their size and functioning can be changed and controlled [50]. Dendrimers have been the subject of extensive research in biocompatibility and their application to a wide range of biological domains, such as drug delivery and immunology, as well as the development of vaccines, antimicrobial agents, and antiviral drugs. Dendrimers belong to a new class of polymers called Dendritic Polymers [51]. These are versatile topologies that are based on conventional linear polymers, crosslinking polymers, branched polymers, and other topologies. The members of the dendrimers family include the bioreducible polymer dendrimer type Gene Delivery.

4.2.9 POLY (ETHYLENIMINE)S

Linear PEI in nature dissolves in ethanol, chloroform, hot water, and environments with low pH. Ethyl ether polymerization to produce branched PEI. By further modifying other polymers, such as poly (2-oxazolines) or N-substituted polyaziridines, linear PEI is made available [52]. Poly (2-ethyl-2-oxazoline) was hydrolyzed to create linear PEI [53, 54].

Lower critical solution temperature (LCST) is observed for poly(N-isopropyl acrylamide) in aqueous solution (PNIPAAm). The 1960s saw research into temperature-responsive polymers [55]. The LCST of thermossensitive poly(N-isopropyl arylamide) has been determined to be 32 C. During phase separation, the heat absorbed and phase diagram separation by the entropy effect have been used to estimate the system's thermodynamic properties [56]. With respect to N-isopropyl-acrylamide, if a single type of monomer is subjected to free radical polymerization, a process known as homo-polymerization results. Radical polymerization frequently uses azobisisobutyronitrile (AIBN) as an initiator. Due to its great landmark in terms of usage medically and biologically, such as the transport of drugs and genes, thermo-responsive polymers have received a lot of interest [56–59].

In the context of temperature fluctuations, the swelling in water of cross-linking poly(N) (N′-N′-altered acrylamide) has been studied. The balance between hydrophilic and hydrophobic properties of the polymer chains, as well as the size, shape, and movement of the alkyl side chains, have both been associated with water swelling thermosensitivity [60]. The approach used to create PNIPAAm-grafted polymers has easily converted the polymer's cell culture surface into hydrophilic and hydrophobic coats [61]. N-isopropyl acrylamide (NIPAAm) and acrylic acid (AAc) were copolymerized to create pH and temperature-sensitive hydrogels [62]. The impact of polyelectrolyte on the LCST of hydrogels that are sensitive to temperature and pH was studied using its swelling ratio pH range. In the same conditions, the percentage of hydrogels that swell in the presence of PAA (poly(allyl amine), also known as polyelectrolytes, was established [63]. It has been briefly explained how the tumor's microenvironment responds to in situ stimuli like pH, redox responsiveness, hypoxia sensitivity, and so forth.

4.2.10 POLY (2-HYDROXYETHYL METHACRYLATE)

A polymer called Poly HEMA, which denoted poly(2-hydroxyethyl methacrylate), produces hydrogel when submerged in another aqueous solution or in water [64]. Poly (PHEMA) hydrogel for intraocular lens material was synthesized by solution polymerization using 2-hydroxyethyl methacrylate (HEMA) as raw material, AIBN, ammonium persulfate or sodium pyrosulfite (APS/SMBS) as catalyst, and ethylene glycol dimethacrylate (EGDMA) or triethylene glycol dimethacrylate (TEGDMA) as crosslinking additive [65]. Using 2-hydroxyethyl methacrylate (HEMA) as raw material, EGDMA or TEGDMA as a crosslinking additive, and ammonium persulfate or sodium pyrosulfite (APS/SMBS) as a catalyst in the solution polymerization process, the poly (PHEMA) hydrogel for intraocular lens material. In order to inhibit cell attachment and promote the development of spheroid, this polymer is often employed to cover cell culture flasks, notably in cancer research. Agar and agarose gels are more established substitutes for pHEMA [66, 67]. TPGDA concentrations for balance swelling, structural characterization, and transport of solutes in swollen poly (HEMA)-crosslinked DDSs were studied

[68]. To elucidate the mechanisms of drug-polymer interactions and their influence on drug-release behavior in controlled-release polymer devices, researchers examined the physicochemical and molecular properties of Pilocarpine derived from poly (HEMA-hydrogels). Biomedical implants frequently use poly (HEMA) hydrogels [69]. Since poly (HEMA) has a high degree of hydrophilicity and is resistant to protein fouling, it is a good choice for coating ventricular catheters [70].

4.2.11 POLYSACCHARIDES

For many years, natural polymers have proven to be effective as medications and their delivery. Owing to the fact that they are non-toxic and biocompatible, polymers that are biodegradable in nature have been extensively researched as competent carrier materials for the delivery of certain drugs. Natural polysaccharides have been researched for use in biomedical and medication delivery industries. The use of modified polymer as a support material for tissue engineering, cell culture, and gene transfer has been discovered. These days, natural polymers have undergone modifications to create innovative biomaterials for applications, including controlled drug delivery. Long chains of carbohydrates called polysaccharides are made up of monosaccharides as monomeric units covalently linked up by glycosidic connections. Using amylase enzymes as a catalyst, this glucose polymer interacts in the presence of water to form monosaccharides or oligosaccharides, which are the component sugars. Natural saccharides typically consist of monosaccharides, which are simple carbohydrates with a n of three or more and a standard formula of $(CH_2O)n$. Monosaccharides include sugars including glucose, fructose, and glyceraldehyde [71]. These organic polymers are employed as biomaterials in medication delivery systems. Glucopyranose monomers have been joined by alpha-connection to form a polymer of glucose popularly known as starch. It is composed of a combination of amylose and amylopectin. Many 100 glucose molecules make up the linear chain of amylose, whereas many 1000 glucose molecules make up the branching molecule known as amylopectin [72].

4.2.12 HYALURONIC ACID, POLY (LACTIC ACID), AND POLY (GLYCOLIC ACID)

Glycolic acid plays a crucial role as an intermediate for the synthesis of organic compounds in a variety of processes, such as polymerization of long-chain esterification and redox reaction. It has been utilized as a monomer during creation of biocompatible copolymers like polyglycolic acid. Two lactic acid molecules have dried out to form the lactone lactide. Lactides polymerize to form biodegradable polyesters called atactic or syndiotactic polylactides in the presence of catalysts [73]. In pharmaceutical technology, lactate and glycolate are created from ordinarily insoluble active substances using lactic and glycolic acids. Additionally, they have discovered that it can be used for its keratolytic and disinfecting qualities in cosmetics, topical preparations, and medication administration [74, 75]. A natural polymer called hyaluronic acid (HA) has the power to target cancer cells that overexpress CD44.

4.2.13 CYCLODEXTRIN DERIVATIVES

A family of cyclic oligosaccharides called cyclodextrin is made up of glucopyranose subunits that are (1, 4) connected. A good molecular chelating agent is cyclodextrin. In nature, cyclodextrins come in three different varieties, that is, 6 units, 7 units, and 8 units β-cyclodextrins. Owing to the cavity size, effectiveness of drug availability, loading and complexation, and reasonably reduced cost of cyclodextrin, it is perfect for drug delivery [76]. 2-hydroxylpropyl derivate, a potent solubilizer with a hydrophilic chain on the exterior and a hydrophobic chain within, is a classical type of cyclodextrin usage in a DDS [77]. They are able to stop drug deterioration, enhance drug stability, and increase drug solubility, all of which increase a drug's bioavailability [78, 79]. For practical applications, such are highly helpful for polymeric DDSs.

4.2.14 ARGININE DERIVATIVES

The amino acid arginine, usually referred to as L-arginine, is employed during the production of proteins [80]. Structurally the molecule possessed a carboxylic acid group, amino group, and a side chain made up of a 3-carbon aliphatic straight chain terminating in a guanidino group. Arginine becomes a charged aliphatic amino acid when the carboxylic acid (COO) is deprotonated at physiological pH, the amino group is protonated (NH_3 $^+$), and the guanidino group is protonated to generate the guanidinium form (C-$(NH_2)^{2+}$).

Arginine adopts an alkaline amino acid structure by deprotonation of its carboxyl group at physiological pH, the amino component and guanidine get protonation to generate NH_3^+ and guanidine form ($C(NH_2)^{2+}$ [81]. A guanidinium group with a pKa of 12.48 caps the distal end of the 3-carbon aliphatic straight chain that makes up the side chain of the amino acid arginine. Therefore, at physiological pH, it is constantly positively charged and protonated. Multiple hydrogen bonds can form in chemical structures due to the displacement of the positive charge due to the double bond conjugation and nitrogen lone pairs [82].

4.2.15 CHITOSAN DERIVATIVES

Chitosan is a cationic polysaccharide synthesized from a natural chitin. It has been widely employed to create polyanions-polyelectrolyte complexes used for the administration of drugs because it is a cationic polymer with advantageous properties [83, 84]. This polysaccharide is popularly known as 2-amino-2-deoxy-d-glucan. It is made of a linear copolymer containing of N-acteyl glucosamine and glucosamine units connected by -(1, 4) links. Synthesized from chitin (2-acetamido-2-deoxy-d-glucan) via deacetylation process. It possesses advantageous biological characteristics such biocompatibility, degradability, mucoadhesiveness, and nontoxicity [85, 86]. The aqueous Chitosan Derivatives like Chitosan Salt, Zwitterionic Chitosan, and Chitosan Oligomers are becoming increasingly popular due to their ability to be dissolved in water for biomedical applications [87–90].

4.3 PHARMACEUTICAL UTILITIES OF GRAFTED POLYMERS

4.3.1 TO ALTER DRUG OF BIOLOGICAL CARRYING CAPACITY

By grafting PLGA with PVA polymer, Dailey et al. [91] increased the capacity of this polymer to carry biomolecules. Sulfobutyl and amino functional groups are attached to the backbone of the PVA, making it easy to create surfaces with either positive or negative surface properties. The grafting process increases hydrophilic properties, which improves the transport of heavier biomolecules, such as DNA, peptides, and proteins. For the use of tetanus toxoid as an oral and nasal vaccine, Jung et al. [92] were the first to introduce a grafted polymer (Sulfobutylated Poly(vinyl Alcohol)-Graft-Poly(Lactide-Co-Glycolide)) associated with tetanus and cholera toxoids, while Ms et al. [93] synthesized and examined polyelectrolyte complex hydrogel microcapsules that are pH-sensitive for the shake oral administration of proteins. Chitosan (CS) grafted alginate hydrogel was created in this study by Omer et al. utilizing the “grafting to” technique, while the coupling agent used was P-benzoquinone, and chitosan chains were clicked onto NH2 groups to form the hydrogel.

4.3.2 TO ACHIEVE TAILORED PHYSICOCHEMICAL PROPERTIES BY GRAFTING OF POLYMERS

Despite the fact that chitosan is an efficient flocculating agent only in acidic conditions, its derivatives, which have chain carboxyl side groups, exhibit zwitterionic properties with high flocculation abilities in acidic and basic media, despite the fact that chitosan is an efficient flocculating agent only in acidic conditions [94]. With the use irradiation, Cai et al. [95] synthesized this hydrogel by grafting chitosan with poly(N-isopropylacrylamide) (PNIPAAm). With this invention, the swelling behavior of these chitosan-g-PNIPAAm hydrogels was improved upon increasing in the number of grafted branches.

Recently, numerous uses of chitosan grafted with acrylic acid as a potential technique of producing hydrophilic and mucoadhesive polymers have been identified. According to Hu et al.'s [96] research, which used silk peptide as a model drug, CS – poly(acrylic acid) (PAA) graft has shown promise as a delivery system for hydrophilic medicines, proteins, and peptides. Through chemical grafting with N-hydroxymethyl acrylamide, acrylamide, and mixtures of 2-acrylamido-2-methylpropane sulfonic acid, and succinic acid treatment, modified psyllium with altered pH-sensitivity, swellability, biodegradability, and ultimately drug release mechanism was created [97].

Thermostable systems were created by Vihola et al. [98] by stabilizing thermally responsive particles with poly (N-vinylcaprolactam) (PVCL), a thermosensitive polymer, and poly (ethylene oxide), a thermoplastic, to create PVCL-graft-C11EO42. These brand-new polymeric substances give hydrogel particles a more thermostable system that can be employed to regulate medication release.

4.3.3 TO ACHIEVE DESIRED DOSAGE FORM CHARACTERISTICS

Poly (lactic-co-glycolic acid), or PLGA, was grafted onto the backbones of water-soluble poly (vinyl alcohol) (PVA) by Pistel et al. [99]. Based on the structure and molecular weight of the copolymer, biodegradable polyesters produced exhibited a significant potential for parenteral DDSs with decreased burst effects and regulated release patterns. Using pH-sensitive interpenetrating polymer network (IPN) beads made of PAAm-g-XG and NaCMC, Kulkarni and Sa [100] created regulated drug delivery devices for ketoprofen. This was accomplished by grafting acrylamide onto the Xanthan Gum (XG) and IPN of PAAm-g-XG- NaCMC.

Transdermal films were created by Patil et al. [101] utilizing modified XG, which was created by grafting acrylamide onto XG. These modified XG films successfully achieved a controlled release of atenolol. N, N-dimethyl-acryl amide, and N-isopropylacryl amide were grafted onto polypropylene film by Conteras-Garcia et al. [102]. These functionalized PP films have excellent bioproperties as well as prolonged norfloxacin release. Siraj et al. [103] combined PVAL with 6-thioguanine and methylcellulose (MC), a natural polymer. This blending produced an IPN microsphere that successfully controlled the release of the medication in the *in vivo* investigation.

4.3.7 TO ACHIEVE SITE-SPECIFIC DELIVERY BY GRAFTING OF POLYMERS

Indomethacin was made as microspheres of polyacrylamide-grafted chitosan that were crosslinked with glutaraldehyde for the treatment of arthritis by Kumbar et al. [104] These microspheres first release indomethacin by a polymer chain relaxing process. However, over extended periods of time, regulated molecular diffusion is primarily responsible for this release.

Other polymeric grafted systems were examined in order to achieve site-specific drug delivery, such as microspheres of grafted chitosan crosslinked with glutaraldehyde that were prepared to encapsulate the antihypertensive drug nifidifine [105] and N-lauryl carboxymethyl chitosan with both hydrophobic and hydrophilic groups that were investigated in relation to the delivery of Taxol to cancerous tissues [106].

Liu et al. [107] created materials based on N-dodecylated chitosan for intracellular delivery of gene therapy. Additionally, the complex was reported to have improved activity and to shield DNA against DNA nuclease. Grafted-chitosan polymeric micelles containing two medicines were created by Nam et al. [108] To create the final construct, HP-TOC-DOX polymeric micelles, the chitosan derivative O-carboxymethyl chitosan was first conjugated with -tocopherol to make -tocopherol O-carboxymethyl (TOC), which was then ligated with doxorubicin and an anti-human epidermal growth factor receptor 2 (HER2) target peptide. According to an author-conducted *in vivo* investigation, the anti-HER targeting peptide improved both cellular uptake and therapeutic efficacy.

4.4 CONCLUSION

Worldwide, pharmaceutical companies are actively working to create new medication delivery methods. IDDSs using polymers are highly significant areas in drug delivery. It offers a number of benefits, including superior swelling capacity, specificity, and mechanical strength, all of which are crucial for the regulated and targeted administration of drugs. One has the chance to create materials with a variety of qualities that will improve the properties and overcome the drawbacks of individual polymer networks by building IDDSs utilizing different polymers. However, if the products have achieved their important goal-clinical use-pharmacologists must be involved in the examination of the PKs and pharmacodynamics.

KEYWORDS

- **biomimetic polymer**
- **drug delivery**
- **graft copolymer**
- **interpenetrating polymer network**
- **low molecular weight**
- **poly (ethylene glycol)**

REFERENCES

1. Din, F., Aman, W., Ullah, I., Qureshi, O. S., Mustapha, O., Shafique, S., & Zeb, A. (2017). Effective use of nano-carriers as drug delivery systems for the treatment of selected tumors. *International Journal of Nanomedicine, 12*, 7291–7309.
2. Tiwari, G., Tiwari, R., Sriwastawa, B., Bhati, L., Pandey, S., Pandey, P., & Bannerjee, S. K. (2012). Drug delivery systems: An updated review. *International Journal of Pharmaceutical Investigations, 2*, 2–11.
3. Sinha, V. R., & Kumria, R. (1998). Bio-absorbable polymers for implantable therapeutic systems. *Drug Development and Industrial Pharmacy, 24*, 1129–1138.

4. Basak, A., Kunduru, K. R., Doppalapudi, S., Domba, A. J., & Khan, W. (2016). Poly (lactic acid) based hydrogels. *Advanced Drug Delivery Reviews, 107*, 192–205. https://doi.org/10.1016/j.addr.2016.07.004.
5. Teleanu, D. M., Chircov, C., Grumezescu, A. M., Volceanov, A., & Teleanu, R. I. (2018). Blood-brain delivery methods using nanotechnology. *Pharmaceutics, 10*(4), 298–305. https://doi.org/10.3390/pharmaceutics10040269.
6. Cacciatore, I., Ciulla, M., Fornasari, E., Marinelli, L., & Di Stefano, A. (2016). Solid lipid nanoparticles as a drug delivery system for the treatment of neurodegenerative diseases. *Expert Opinion on Drug Delivery, 13*(8), 1121–1131. https://doi.org/10.1080/17425247.2016.1196767.
7. Patel, M., Souto, E. B., & Singh, K. K. (2013). Advances in brain drug targeting and delivery: Limitations and challenges of solid lipid nanoparticles. *Expert Opinion on Drug Delivery, 10*, 889–905.
8. Kumar, M. N. V. R., & Kumar, N. (2001). Polymeric controlled drug delivery systems: Perspective issues and opportunities. *Drug Development and Industrial Pharmacy, 27*(1), 1–30.
9. Blume, G., & Cevc, G. (1990). Liposomes for the sustained drug release *in vivo*. *Biochimica et Biophysica Acta, 1029*(1), 91–97.
10. Liechty, W. B., Kryscio, D. R., Slaughter, B. V., & Peppas, N. A. (2010). Polymers for drug delivery systems. *Annual Review of Chemical and Biomolecular Engineering, 1*, 149–173.
11. Giusti, P., Lazzeri, L., & Cascone, M. G. (1996). Bioartificial materials. In J. C. Salomone (Ed.), *The Polymeric Materials Encyclopedia* (pp. 538–549). CRC Press.
12. Cascone, M. G. (1997). Dynamic-mechanical properties of bioartificial polymeric materials. *Polymer International, 43*(1), 55–69.
13. Sperling, L. H., & Hu, R. (2003). Interpenetrating polymer networks. In L. A. Utracki (Ed.), *Polymer Blends Handbook* (pp. 417–447). Springer.
14. Ignat, L., & Stanciu, A. (2003). Advanced polymers: Interpenetrating polymer networks. In A. K. Kulshreshtha & C. Vasile (Eds.), *Handbook of Polymer Blends and Composites* (Vol. 3, pp. 275–280). Rapra Technology.
15. Kudela, V. (1987). Hydrogels. In J. I. Kroschwitz (Ed.), *Encyclopedia of Polymer Science and Engineering* (pp. 783–807). Wiley.
16. Work, W. J., Horie, K., Hess, M., & Stepto, R. F. T. (2004). Definitions of terms related to polymer blends, composites, and multiphase polymeric materials. *Pure and Applied Chemistry, 76*(11), 1985–2007.
17. Aylsworth, J. W. (1914). Plastic composition. *US Patent 1111284*. September 1914.
18. Millar, J. R. (1960). Interpenetrating polymer networks, styrene-divinyl benzene copolymer with two and three interpenetrating networks and their sulphonates. *Journal of the Chemical Society*, 1311–1319.
19. Ou, M., Wang, X. L., Xu, R., Chang, C. W., Bull, D. A., & Kim, S. W. (2008). Novel biodegradable poly (disulfide amine) s for gene delivery with high efficiency and low cytotoxicity. *Bioconjugate Chemistry, 19*, 626–633.
20. Nam, J. P., Kim, S., & Kim, S. W. (2018). Design of PEI-conjugated bio-reducible polymer for efficient gene delivery. *International Journal of Pharmaceutics, 545*, 295–305. https://doi.org/10.1016/j.ijpharm.2018.04.051.

21. Lee, Y. S., & Kim, S. W. (2014). Bioreducible polymers for therapeutic gene delivery. *Journal of Controlled Release, 190*, 424–439. https://doi.org/10.1016/j.jconrel.2014.04.012.
22. Rai, R., Alwani, S., & Badea, I. (2019). Polymeric nanoparticles in gene therapy: New avenues of design and optimization for delivery applications. *Polymers (Basel), 11*(4), 745–749. https://doi.org/10.3390/polym11040745.
23. Kim, T., & Kim, S. W. (2011). Bioreducible polymers for gene delivery. *Reactive & Functional Polymers, 71*, 344–349. https://doi.org/10.1016/j.reactfunctpolym.2010.11.016.
24. Wen, Y., Pan, S., Luo, X., Zhang, X., Zhang, W., & Feng, M. (2009). A biodegradable low molecular weight polyethylenimine derivative as a low toxicity and efficient gene vector. *Bioconjugate Chemistry, 20*, 322–332. https://doi.org/10.1021/bc800428y.
25. Wen, Y., Pan, S., Luo, X., Zhang, W., Shen, Y., & Feng, M. (2010). PEG- and PDMAEG-graft-modified branched PEI as a novel gene vector: Synthesis, characterization, and gene transfection. *Journal of Biomaterials Science, Polymer Edition, 21*, 1103–1126.
26. Kim, S. W., Nam, J. P., Kim, S., & Sung, Y. K. (2018) Recent development of bio-reducible polymers for efficient gene delivery systems. *Journal of Cancer Treatment and Diagnosis, 2*, 17–23.
27. Nam, J. P., Park, J. K., Son, D. H., Kim, T. H., Park, S. J., Park, S. C., Choi, C., Jang, M. K., & Nah, J. W. (2014). Evaluation of polyethylene glycol conjugated novel polymeric antitumor drug for cancer therapy. *Colloids and Surfaces B: Bio Interfaces, 120*, 168–175.
28. Nam, K., Jung, S., Nam, J. P., & Kim, S. W. (2015). Poly (ethylenimine) conjugated bioreducible dendrimer for efficient gene delivery. *Journal of Controlled Release, 220*, 447–455. https://doi.org/10.1016/j.jconrel.2015.11.005.
29. Sung, Y. K., & Kim, S. W. (2019). Recent advances in the development of gene delivery systems. *Biomaterials Research, 23*, 8–12. https://doi.org/10.1186/s40824-019-0156-z.
30. Zhong, Z., Feijen, J., Lok, M. C., Hennink, W. E., Christensen, L. V., Yockman, J. W., Kim, Y. H., & Kim, S. W. (2005). Low molecular weight linear polyethylenimine-b-poly (ethylene glycol)-b-polyethylenimine triblock copolymers: Synthesis, characterization, and *in vitro* gene transfer properties. *Biomacromolecules, 6*, 3440–3448.
31. Park, M. R., Han, K. O., Han, I. K., Cho, M. H., Nah, J. W., Choi, Y. J., & Cho, C. S. (2005). Degradable polyethylenimine-alt-poly (ethylene glycol) copolymers as novel gene carriers. *Journal of Controlled Release, 105*, 367–380.
32. Bagalkot, V., Farokhzad, O. C., Langer, R., & Jon, S. (2006). An aptamer–doxorubicin physical conjugate as a novel targeted drug-delivery platform. *Angewandte Chemie International Edition, 45*, 8149–8152.
33. Speck, O., Speck, D., Horn, R., Gantner, J., & Sedlbauer, K. P. (2017). Biomimetic bio-inspired biomorph sustainable. An attempt to classify and clarify biology-derived technical developments. *Bioinspiration & Biomimetics, 12*(1), 011004. https://doi.org/10.1088/1748-3190/12/1/011004.
34. Sabua, C., Rejob, C., Kottab, S., & Pramoda, K. (2018). Bioinspired and biomimetic systems for advanced drug and gene delivery. *Journal of Controlled Release, 287*, 142–155. https://doi.org/10.1016/j.jconrel.2018.08.033.
35. Safari, J., & Zarnegar, Z. (2014). Advanced drug delivery systems: Nanotechnology of health design – A review. *Journal of the Saudi Chemical Society, 18*, 85–99.

36. Li, Y., Thambi, T., & Lee, D. S. (2018). Co-delivery of drugs and genes using polymeric nanoparticles for synergistic cancer therapeutic effects. *Advanced Healthcare Materials, 7*, 1700886. https://doi.org/10.1002/adhm.201700886.
37. Yoo, J. W., Irvine, D. J., Discher, D. E., & Mitragotri, S. (2011). Bio-inspired, bioengineered and biomimetic drug delivery carriers. *Nature Reviews Drug Discovery, 10*, 521–535. https://doi.org/10.1038/nrd3499.
38. Yang, L., Li, J., & Kopeček, J. (2019). Biorecognition: A key to drug-free macromolecular therapeutics. *Biomaterials, 190*, 11–23.
39. Törmälä, P., Pohjonen, T., & Rokkanen, P. (1998). Bio-absorbable polymers: Materials technology and surgical applications. *Proceedings of the Institution of Mechanical Engineers, Part H: Journal of Engineering in Medicine, 212*, 101–111. https://doi.org/10.1243/0954411981533872.
40. Pulapura, S., & Kohn, J. (1992). Trends in the development of bio-resorbable polymers for medical applications. *Journal of Biomaterials Applications, 6*, 216–250.
41. Lee, T. S., & Bee, S. T. (2019). *Polylactic Acid (Second Edition): A Practical Guide for the Processing, Manufacturing and Applications of PLA*. Plastics Design Library. Elsevier B.V. pp. 53–95.
42. Heller, J., Barr, J., YNg, S., Abdellauoi, K. S., & Gurny, R. (2002). Poly (ortho esters): Synthesis, characterization, properties and uses. *Advanced Drug Delivery Reviews, 54*(10), 1015–1039. https://doi.org/10.1016/S0169-409X(02)00055-8.
43. Kumar, N., Langer, R. S., & Domb, A. J. (2002). Polyanhydrides: An overview. *Advanced Drug Delivery Reviews, 54*(8), 889–910. https://doi.org/10.1016/S0169-409X(02)00050-9.
44. Kamaly, N., Yameen, B., Wu, J., & Farokhzad, O. C. (2016). Degradable controlled-release polymers and polymeric nanoparticles: Mechanisms of controlling drug release. *Chemical Reviews, 116*(4), 2602–2633.
45. Nicolas, J., Mura, S., Brambilla, D., Mackiewicz, N., & Couvreur, P. (2013). Design, functionalization strategies and biomedical applications of targeted biodegradable/biocompatible polymer-based nanocarriers for drug delivery. *Chemical Society Reviews, 42*(3), 1147–1215. https://doi.org/10.1039/c2cs35265f.
46. Yang, J., Zhang, R., Pan, H., Li, Y., Fang, Y., Zhang, L., & Kopeček, J. (2017). Backbone degradable N-(2-hydroxypropyl) methacrylamide copolymer conjugates with gemcitabine and paclitaxel: Impact of molecular weight on activity toward human ovarian carcinoma xenografts. *Molecular Pharmaceutics, 14*(4), 1384–1394. https://doi.org/10.1021/acs.molpharmaceut.6b01005.
47. Zhang, R., Yang, J., Zhou, Y., Shami, P. J., & Kopeček, J. (2016). N-(2-hydroxypropyl) methacrylamide copolymer–drug conjugates for combination chemotherapy of acute myeloid leukemia. *Macromolecular Bioscience, 16*(1), 121–128.
48. Zhang, L., Zhang, R., Yang, J., Wang, J., & Kopeček, J. (2016). Indium-based and iodine-based labeling of HPMA copolymer-epirubicin conjugates: Impact of structure on the *in vivo* fate. *Journal of Controlled Release, 240*, 306–318.
49. Gillies, E. R., & Fréchet, J. M. J. (2005). Dendrimers and dendritic polymers in drug delivery. *Drug Discovery Today, 10*(1), 35–43.
50. Menjoge, A. R., Kannan, R. M., & Tomalia, D. A. (2010). Dendrimer-based drug and imaging conjugates: Design considerations for nano-medical applications. *Drug Discovery Today, 15*(5–6), 171–185.

51. Nam, H. Y., Nam, K., Lee, M., Kim, S. W., & Bull, D. A. (2012). Dendrimer type bio-reducible polymer for efficient gene delivery. *Journal of Controlled Release, 160*(3), 592–600. https://doi.org/10.1016/j.jconrel.2012.04.025.
52. Tanaka, R., Ueoka, I., Takaki, Y., Kataoka, K., & Saito, S. (1983). High molecular weight linear polyethylenimine and poly(N-methylethylenimine). *Macromolecules, 16*(6), 849–853. https://doi.org/10.1021/ma00240a003.
53. Weyts, K. F., & Goethals, E. J. (1988). New synthesis of linear polyethyleneimine. *Polymer Bulletin, 19*(1), 13–19. https://doi.org/10.1007/bf00255018.
54. Brissault, B., Kichler, A., Guis, C., Leborgne, C., Danos, O., & Cheradame, H. (2003). Synthesis of linear polyethylenimine derivatives for DNA transfection. *Bioconjugate Chemistry, 14*(3), 581–587. https://doi.org/10.1021/bc0200529.
55. Heskins, M., & Guillet, J. E. (1968). Solution properties of poly (N-isopropyl acrylamide). *Journal of Macromolecular Science, Part A: Chemistry, 2*(6), 1441–1455.
56. Zheng, L., Qiulin, L., Duanguang, Y., Yong, G., & Xujun, L. (2013). Well-defined poly(N-isopropylacrylamide) with a bifunctional end-group: Synthesis, characterization, and thermoresponsive properties. *Designed Monomers and Polymers, 16*(5), 465–474. https://doi.org/10.1080/15685551.2012.747165.
57. Schmaljohann, D. (2006). Thermo- and pH-responsive polymers in drug delivery. *Advanced Drug Delivery Reviews, 58*(15), 1655–1670.
58. Ma, Y., Hou, S., Ji, B., Yao, Y., & Feng, X. (2010). A novel temperature-responsive polymer as a gene vector. *Macromolecular Bioscience, 10*(2), 202–210.
59. Weber, C., Richard, H., & Schubert, U. S. (2012). Temperature responsive bio-compatible polymers based on poly (ethylene oxide) and poly (2-oxazoline) s. *Progress in Polymer Science, 37*(5), 686–714.
60. Bae, Y. H., Okano, T., & Kim, S. W. (1990). Temperature dependence of swelling of crosslinked poly (N, N′-alkyl substituted acrylamides) in water. *Journal of Polymer Science Part B: Polymer Physics, 28*(6), 923–936. https://doi.org/10.1002/polb.1990.090280609.
61. Okano, T., Yamada, N., Sakai, H., & Sakurai, Y. (1993). A novel recovery system for cultured cells using plasma-treated polystyrene dishes grafted with poly (N-isopropyl acrylamide). *Journal of Biomedical Materials Research, 27*(9), 1243–1251. https://doi.org/10.1002/jbm.820271005.
62. Yoo, M. K., Sung, Y. K., Lee, Y. M., & Cho, C. S. (2000). Effect of polyelectrolyte on the lower critical solution temperature of poly (N-isopropyl acrylamide) in the poly (NIPAAm-co-acrylic acid) hydrogel. *Polymer, 41*(15), 5713–5719. https://doi.org/10.1016/S0032-3861(99)00779-2.
63. Zhang, Q., Ko, N. R., & Oh, I. K. (2012). Recent advances in stimuli-responsive degradable block copolymer micelles: Synthesis and controlled drug delivery applications. *Chemical Communications, 48*(57), 7542–7552. https://doi.org/10.1039/c2cc32408c.
64. Wichterle, O., & Lím, D. (1960). Hydrophilic gels for biological use. *Nature, 185*(4716), 117–118. https://doi.org/10.1038/185117a0.
65. Sung, Y. K., Jhon, M. S., Gregonis, D. E., & Andrade, J. D. (1981). Thermal and pulse nuclear magnetic resonance analysis of water in poly (2-hydroxyethyl methacrylate). *Journal of Applied Polymer Science, 26*(11), 3719–3726.
66. Katt, M. E., Placone, A. L., Wong, A. D., Xu, Z. S., & Searson, P. C. (2016). *In vitro* tumor models: Advantages, disadvantages, variables, and selecting the right platform.

Frontiers in Bioengineering and Biotechnology, 4, 12–18. https://doi.org/10.3389/fbioe.2016.00012.
67. Friedrich, J. S., Claudia, E. R., & Kunz-Schughart, L. A. (2009). Spheroid-based drug screen: Considerations and practical approach. *Nature Protocols, 4*(3), 309–324. https://doi.org/10.1038/nprot.2008.226.
68. Ferreira, L., Vidal, M. M., & Gil, M. H. (2000). Evaluation of poly (2-hydroxyethyl methacrylate) gels as drug delivery systems at different pH values. *International Journal of Pharmaceutics, 194*(2), 169–180. https://doi.org/10.1016/S0378-5173(99)00375-0.
69. Hsiue, G. H., & Cheng, C. C. (2001). Poly (2-hydroxyethyl methacrylate) film as a drug delivery system for Pilocarpine. *Biomaterials, 22*(16), 1763–1769. https://doi.org/10.1016/S0142-9612(00)00336-7.
70. Hanak, B. W., Hsieh, C. Y., Donaldson, W., Browd, S. R., Lau, K. S., & Shain, W. (2018). Reduced cell attachment to poly (2-hydroxyethyl methacrylate)-coated ventricular catheters *in vitro*. *Journal of Biomedical Materials Research Part B: Applied Biomaterials, 106*(5), 1268–1279.
71. Matthews, C. E., Van Holde, K. E., & Ahern, K. G. (1999). *Biochemistry* (3rd ed.). Benjamin Cummings. ISBN 0-8053-3066-6.
72. Green, M. M., Blankenhorn, G., & Hart, H. (1975). Which starch fraction is water-soluble, amylose or amylopectin? *Journal of Chemical Education, 52*(11), 729–738.
73. Odile, D. C., Blanca, V., & Didier, B. (2004). Controlled ring-opening polymerization of lactide and glycolide. *Chemical Reviews, 104*(7), 6147–6176. https://doi.org/10.1021/cr0203571.
74. Yoo, D. K., Kim, D., & Lee, D. S. (2006). Synthesis of lactide from oligomeric PLA: Effects of temperature, pressure, and catalyst. *Macromolecular Research, 14*(6), 510–516.
75. Pourasghar, M., Koenneke, A., Meiers, P., & Schneider, M. (2019). Development of a fast and precise method for simultaneous quantification of the PLGA monomers lactic and glycolic acid by HPLC. *Journal of Pharmaceutical Analysis, 9*(1), 100–107.
76. Karande, P., & Mitragotri, S. (2009). Enhancement of transdermal drug delivery via synergistic action of chemicals. *Biochimica et Biophysica Acta, 1788*(11), 2362–2373. https://doi.org/10.1016/j.bbamem.2009.08.015.
77. Manosroi, J., Apriyani, M. G., Foe, K., & Manosroi, A. (2005). Enhancement of the release of azelaic acid through the synthetic membranes by inclusion complex formation with hydroxypropyl-beta-cyclodextrin. *International Journal of Pharmaceutics, 293*(1–2), 235–240. https://doi.org/10.1016/j.ijpharm.2005.01.009.
78. Wang, S., Tan, M., Zhong, Z., Chen, M., & Wang, Y. (2011). Nanotechnologies for curcumin: An ancient puzzler meets modern solutions. *Journal of Nanomaterials, 2011*, 1–8. https://doi.org/10.1155/2011/723178.
79. Martin, E. M., & Valle, D. (2004). Cyclodextrins and their uses: A review. *Process Biochemistry, 39*(7), 1033–1046.
80. Tapiero, H., Mathé, G., Couvreur, P., & Tew, K. D. (2002). L-arginine. *Biomedicine & Pharmacotherapy, 56*(9), 439–445. https://doi.org/10.1016/s0753-3322(02)00284-6.
81. Wu, G., Jaeger, L. A., Bazer, F. W., & Rhoads, J. M. (2004). Arginine deficiency in preterm infants: Biochemical mechanisms and nutritional implications. *Journal of Nutritional Biochemistry, 15*(8), 442–451.
82. Skipper, A. (1998). *Dietitian's Handbook of Enteral and Parenteral Nutrition*. Jones & Bartlett Learning.

83. Wu, Q. X., Lin, D. Q., & Yao, S. J. (2014). Design of chitosan and its water-soluble derivatives-based drug carriers with polyelectrolyte complexes. *Marine Drugs, 12*(12), 6236–6253. https://doi.org/10.3390/md12126236.
84. Hamman, J. H. (2010). Chitosan-based polyelectrolyte complexes as potential carrier materials in drug delivery systems. *Marine Drugs, 8*(5), 1305–1322. https://doi.org/10.3390/md8051305.
85. Onishi, H., & Machida, Y. (1999). Biodegradation and distribution of water-soluble chitosan in mice. *Biomaterials, 20*(2), 175–182.
86. Zhang, H., Alsarra, I. A., & Neau, S. H. (2002). An *in vitro* evaluation of a chitosan-containing multi-particulate system for macromolecule delivery to the colon. *International Journal of Pharmaceutics, 239*(2), 197–205.
87. Xu, P. S., Bajaj, G., Shugg, T., van Alstine, W. G., & Yeo, Y. (2010). Zwitterionic chitosan derivatives for pH-sensitive stealth coating. *Biomacromolecules, 11*(9), 2352–2358.
88. Bajaj, G., van Alstine, W. G., & Yeo, Y. (2012). Zwitterionic chitosan derivative, a new biocompatible pharmaceutical excipient, prevents endotoxin-mediated cytokine release. *PLoS ONE, 7*(4), e34516.
89. Čalija, B., Cekić, N., Savić, S., Daniels, R., Marković, B., & Milić, J. (2013). pH-sensitive microparticles for oral drug delivery based on alginate/oligo-chitosan/Eudragit® L100–55 "sandwich" polyelectrolyte complex. *Colloids and Surfaces B: Biointerfaces, 110*, 395–402.
90. Luo, Y., & Wang, Q. (2014). Recent development of chitosan-based polyelectrolyte complexes with natural polysaccharides for drug delivery. *International Journal of Biological Macromolecules, 64*, 353–356.
91. Dailey, L. A., Wittmar, M., & Kissel, T. (2005). The role of branched polyesters and their modifications in the development of modern drug delivery vehicles. *Journal of Controlled Release, 101*(1–3), 137–149.
92. Jung, T., Kamm, W., Breitenbach, A., Hungerer, K. D., Hunct, E., & Kissel, T. (2001). Tetanus toxoid loaded nanoparticles from sulfobutylated poly (vinyl alcohol)-graft-poly(lactide-co-glycolide): Evaluation of antibody response after oral and nasal application in mice. *Pharmaceutical Research, 18*(3), 352–360.
93. Ms, M. E., Am, O., Ma, W., Tm, T., Ms, A. B., & Sa, I. (2015). Novel smart pH sensitive chitosan grafted alginate hydrogel microcapsules for oral protein delivery: I. Preparation and characterization. *International Journal of Pharmacy and Pharmaceutical Sciences, 7*(10), 320–326.
94. Bashir, A., Warsi, M. H., & Sharma, P. K. (2016). An overview of natural gums as pharmaceutical excipients: Their chemical modification. *World Journal of Pharmaceutical and Medical Research, 5*(4), 2025–2039.
95. Cai, H., Pu, Z., Chuan, P., Lin, B., & Xia, X. (2005). Synthesis and characterization of thermo-and pH-sensitive hydrogels based on chitosan-grafted N-isopropylacrylamide via γ-radiation. *Radiation Physics and Chemistry, 74*(1), 26–30.
96. Hu, Y., Jiang, X., Ding, Y., Ge, H., Yuan, Y., & Yang, C. (2002). Synthesis and characterization of chitosan-poly (acrylic acid) nanoparticles. *Biomaterials, 23*(15), 3193–3201.
97. Kumar, R., & Sharma, K. (2013). Biodegradable polymethacrylic acid grafted psyllium for controlled drug delivery systems. *Frontiers in Chemical Science and Engineering, 7*(1), 116–122.

98. Vihola, H., Laukkanen, A., Tenhu, H., & Hirvonen, J. (2008). Drug release characteristics of physically crosslinked thermosensitive poly(N-vinylcaprolactam) hydrogel particles. *Journal of Pharmaceutical Sciences, 97*(11), 4783–4793.
99. Pistel, K. F., Breitenbach, A., Zange-Volland, R., & Kissel, T. (2001). Brush-like branched biodegradable polyesters, Part III: Protein release from microspheres of poly (vinyl alcohol)-graft-poly (D, L-lactic-co-glycolic acid). *Journal of Controlled Release, 73*(1), 7–20.
100. Kulkarni, R. V., & Sa, B. (2008). Evaluation of pH-sensitivity and drug release characteristics of polyacrylamide-grafted-xanthan-carboxymethyl cellulose-based pH-sensitive interpenetrating network hydrogel beads. *Drug Development and Industrial Pharmacy, 34*(12), 1406–1414.
101. Patil, S. A. (2007). Evaluation and controlled release characteristics of modified xanthan films for transdermal delivery of atenolol. *Drug Development and Industrial Pharmacy, 33*(1), 79–90.
102. Contreras-García, A., Bucio, E., Concheiro, A., Álvarez-Lorenzo, C., Bucio, E., Concheiro, A., & Contreras-García, A. (2011). *Journal of Bioactive and Compatible Polymers, 26*(4), 405–419.
103. Siraj, S., Sudhakar, P., Rao, U. S., Sekharnath, K. V., Rao, K. C., & Subha, M. C. (2014). Interpenetrating polymer network microspheres of poly (vinyl alcohol)/methyl cellulose for controlled release studies of 6-thioguanine. *International Journal of Pharmaceutics and Pharmaceutical Sciences, 6*(9), 101–106.
104. Kumbar, S. G., Soppimath, K. S., & Aminabhavi, T. M. (2002). Synthesis and characterization of polyacrylamide-grafted chitosan hydrogel microspheres for the controlled release of indomethacin. *Journal of Applied Polymer Science, 87*(11), 1525–1536.
105. Yoshioka, H., Nonaka, K., Fukuda, K., & Kazama, S. (1995). Chitosan-derived polymer-surfactants and their micellar properties. *Bioscience, Biotechnology, and Biochemistry, 59*(10), 1901–1904.
106. Alves, N. M., & Mano, J. F. (2008). Chitosan derivatives obtained by chemical modifications for biomedical and environmental applications. *International Journal of Biological Macromolecules, 43*(5), 401–414.
107. Liu, W. E., Yao, K. D., & Liu, Q. G. (2001). Formation of a DNA/N-dodecylated chitosan complex and salt-induced gene delivery. *Journal of Applied Polymer Science, 82*(13), 3391–3395.
108. Nam, J. P., Lee, K. J., Choi, J. W., Yun, C. O., & Nah, W. W. (2015). Targeting delivery of tocopherol and doxorubicin grafted-chitosan polymeric micelles for cancer therapy: *In vitro* and *in vivo* evaluation. *Colloids and Surfaces B: Biointerfaces, 133*, 254–262.

CHAPTER 5

Grafted Polysaccharide for the Application in Gene Delivery

AJAYKUMAR CHITTIPOLU,[1] AKBAR PASHA,[2] DONETI RAVINDAR,[2] and CHEMBULI ABHISHEK[3]

[1]*Vision College of Pharmaceutical Sciences and Research, JNTUH, Hyderabad, Telangana, India*

[2]*Department of Genetics and Biotechnology, Osmania University, Hyderabad, Telangana, India*

[3]*Department of Biotechnology, North-Eastern University, USA*

ABSTRACT

Plant polysaccharides and animal-based protein polymers are widely used for a variety of biomedical applications, including drug delivery and tissue regeneration, recognition, and acceptability of the polymers with extracellular matrix components of humans due to the biochemical similarity. Since they are less expensive, less toxic, and more readily available than their synthetic equivalents, polysaccharides have an advantage over synthetic polymers. Because they are derived from renewable raw materials, polysaccharides seem to be a highly attractive replacement option. Natural polysaccharides contain a variety of special qualities, but they also have significant drawbacks, most notably a short shelf life and a propensity for bacterial invasion, which limits their potential as durable materials. Certain synthetic polymers are readily tailor-made and biodegradable. Hence, higher shelf life and biodegradability can be achieved by combining the unique benefits of polysaccharides and synthetic polymers. Grafting is one of the design strategies that have attracted attention to improve the compatibility

Graft Copolymers of Polysaccharides for Drug Delivery Systems.
Rishabha Malviya and Sonali Sundram (Eds.)

of synthetic and natural polymers. Graft copolymerization is an important method for enhancing the innovative characteristics of backbone polymers. It is a chemical process that alters the desired properties of natural fibers without changing their basic behavior. The grafting concept enables contact with anionic surfaces of double-strand DNA strands in three dimensions. The polysaccharide-based gene delivery technology suggests that the cationic polysaccharide's structure has a vital impact on the transfection activity for gene delivery. The current chapter focuses on the application of polysaccharide-based materials in gene-based delivery for regenerative engineering.

5.1 INTRODUCTION

Gene therapy directs the underlying genetic issue to cure or prevent illness by changing a person's genetic composition. The potential of the technology to significantly advance medical practice has been ever since the discovery of recombinant DNA technology. Most technological developments that have had an impact on clinical patient management have involved the creation of novel molecular diagnostic methods. Recombinant DNA technology has also made it possible to define the roles played by gene products in the etiology of human illness. Clinical therapies have already become more accurate and successful because of their capacity to describe illness in such molecular terms. The concept behind gene therapy, however, is that a patient's unique cells may transfer genetic material into a patient's condition instead of using traditional medications. Some scientists and medical professionals are looking into gene therapy techniques, such as gene knockout, mutation, and replacement in genetic material. Several diseases, including cancer, genetic disorders, and some viral infections, may be treated with gene therapy, but the method is risky and must first be made safe before it can be used effectively. Gene therapy has only been investigated for ailments for which there are currently no other available treatments [1]. To start gene expression, genetic molecules enter the nucleus of the host cell. Gene therapy has evolved to provide genetic information so that a patient's somatic cells can create specific therapeutic proteins to alter hereditary illnesses [2].

Successful gene delivery design requires an understanding of the interactions between targeted cellular components and the delivery mechanism. Three fundamental systems are central to gene delivery: (i) plasmid gene expression, which regulates gene activity, (ii) therapeutic gene encoding, which produces specific proteins, and (iii) plasmid-based gene expression targeting specific sites in the body [3–5]. For gene delivery to be effective, the foreign genetic material must maintain stability throughout the process. Two main vector-based methods are commonly used for delivering genetic material: viral and non-viral systems. Viral vectors, which were among the first methods used for transgenic expression, include human foamy virus (HFV), lentivirus, retrovirus, poxvirus, herpesvirus, adeno-associated virus, and adenovirus. These vectors are effective for gene therapy, often modified by deleting specific regions to limit replication and increase safety. However, viral vectors have drawbacks, such as high immunogenicity, which can provoke inflammatory responses, damage transduced tissues, and cause toxicity.

Non-viral gene delivery methods were the first to document cellular phenotypic changes triggered by DNA exposure. Unlike viral systems, non-viral systems include both chemical techniques, such as cationic liposomes and polymers, and physical techniques like gene guns, electroporation, particle bombardment, ultrasound, and magnetofection. Although non-viral methods are generally less efficient at delivering genes compared to viral systems, they offer advantages in terms of lower cost, accessibility, and reduced immune response activation [5–8].

Recent advancements in gene delivery system design have included both viral and non-viral approaches. Non-viral gene delivery methods encompass physical and chemical techniques. Physical methods, such as non-viral gene transfer, are used to introduce genetic material across cell membranes. Examples include hydroporation, magnetofection, sonoporation, photoporation, and ballistic DNA injection [8]. "Needle injection" refers to the direct infusion of genetic material, while ballistic DNA injection delivers gold-coated DNA fragments into cells. Electroporation creates pores in the cell membrane, allowing genetic material to enter. Sonoporation uses sound waves to open cell membrane pores, and photoporation employs laser pulses to achieve the same effect. Magnetofection uses magnetic particles complexed with DNA to concentrate nucleic acid particles in target cells, while hydroporation alters cell permeability through a hydrodynamic capillary effect [9]. Non-viral chemical delivery systems involve the creation of particles that aid in gene delivery using artificial or organic materials. Polymers and liposomes are the two primary types of non-viral vectors. Liposome-based vectors enable gene transfer by forming lipoplexes, which occur when positively charged liposomes interact with negatively charged

DNA. Similarly, non-viral polymer-based vectors combine with DNA to form polyplexes, facilitating gene delivery [8].

5.2 POLYMERS

In Greek, the terms poly, which represents "many," and meros, which represents "parts or units of high molecular mass," are combined to form the English word "polymer." Every molecule is made up of a colossal number of distinct structural components that are regularly connected to one another by covalent bonds. By connecting concurrently, numerous similar component molecules (monomers) can become a polymer; polymers are enormous macromolecules with a high molecular weight that are known as macromolecules. Polymerization (a process of monomers connected to produce polymers) [10].

Organic polymers are naturally occurring materials derived from sources such as plants and animals, including proteins, cellulose, starch, and resins. Semi-artificial polymers are created by chemically modifying natural polymers like starch and silicones to alter their physical properties. Artificial, or synthetic, polymers are produced through laboratory polymerization of basic chemical compounds to create fibers such as nylon, polyethylene, polystyrene, synthetic rubber, PVC, and Teflon.

Natural polymers are desirable due to their lower cost, availability, biodegradability, and potential biocompatibility. They hold significant potential in the pharmaceutical, food, and cosmetic industries. As biogenic materials, natural polymers are ideal for drug delivery systems due to their biological characteristics, such as the ability to recognize and interact with cells, be broken down by enzymes, mimic the extracellular matrix, and offer chemical flexibility.

5.2.1 ORIGIN OF NATURAL POLYMER

Natural sources are numerous, renewable, and easy to access, which is probably one of the reasons why natural polymers have attracted interest. Numerous beneficial characteristics of monosaccharide polymers include their high stability, nontoxicity, hydrophilicity, biodegradability, and ability to form gels. Polysaccharides are widespread natural biopolymers and have special biological, physical, and chemical characteristics. The polymers are distinguished by the variety and complexity of the polysaccharides. Two distinct structural characteristics of polysaccharides—first, the ability of monosaccharides to connect to one another in many ways and in a configuration, and second, the existence of a branched sidechain can explain their complexity and variety [11, 12].

Both natural and synthetic polymers have found use in pharmaceutical research due to their distinct capacities to control and release the behavior of various drugs and genes. The efficacy of intracellular gene delivery is greatly influenced by the shape, size, and surface charge of polymeric nanoparticles since these characteristics are critical in deciding the fate of genes. Due to their extraordinary physical and biological characteristics, as well as the numerous reactive groups that are capable of undergoing chemical modifications, polysaccharides have gained greater interest in this respect [12, 13]. Polysaccharides are made up of monosaccharides that are frequently bonded together by O-glycosidic linkages. There are several sources of polysaccharides (chitosan, hyaluronan, and chondroitin), including cellulose, pectin, GG, and the two polysaccharides found in algae, alginate, and carrageenan [14, 15]. Depending on structures, polysaccharides demonstrate a variety of physicochemical traits and have a wide range of functions. For some applications, it is simple to alter a variety of polysaccharides, including those with amino, carboxyl, and hydroxyl. Since changing the backbone of a polysaccharide polymer significantly alters its pharmacological and pharmacokinetic (PK) properties, determining the efficacy of polysaccharide-based drug carriers is essential [15–18].

5.2.2 GRAFTING POLYSACCHARIDES

Natural polysaccharides possess several valuable qualities, but they also have significant drawbacks, such as a short shelf life and susceptibility to bacterial invasion, which limits their durability. Some synthetic polymers, on the other hand, are easily customized and biodegradable. By combining the unique benefits of polysaccharides and synthetic polymers, it is possible to achieve both enhanced shelf life and biodegradability. Starch derivatives, for example, are fully biodegradable polymers with potential applications in the biomedical and environmental fields. Polysaccharides, proteins, and nucleic acids are the fundamental building blocks of organic systems. To meet the needs of industry and science, biopolymers that mimic synthetic polymers hold significant promise for future research [19]. Synthetic polymers exhibit unique chemical properties and are used in various fields, including stimuli-responsive polymers. The grafting of synthetic polymers onto biomedical and biodegradable materials has garnered significant interest from researchers [20]. Graft copolymerization is a key method for enhancing the properties of natural polymers by introducing new characteristics without altering their inherent behavior [21]. Many researchers have successfully grafted various polysaccharide backbones, such as starch, chitosan, and cellulose, using vinyl monomers, binary monomeric mixes, and radiation methods [22]. Graft copolymerization with suitable monomers is one of the most effective modification techniques, allowing for controlled attributes of the grafted product. This process molecularly tailors the final macromolecules for specific applications. Vinyl monomer graft copolymerization is a widely used, practical, and cost-effective method for chemically modifying natural polymers [23].

5.3 CLASSIFICATION OF POLYSACCHARIDES

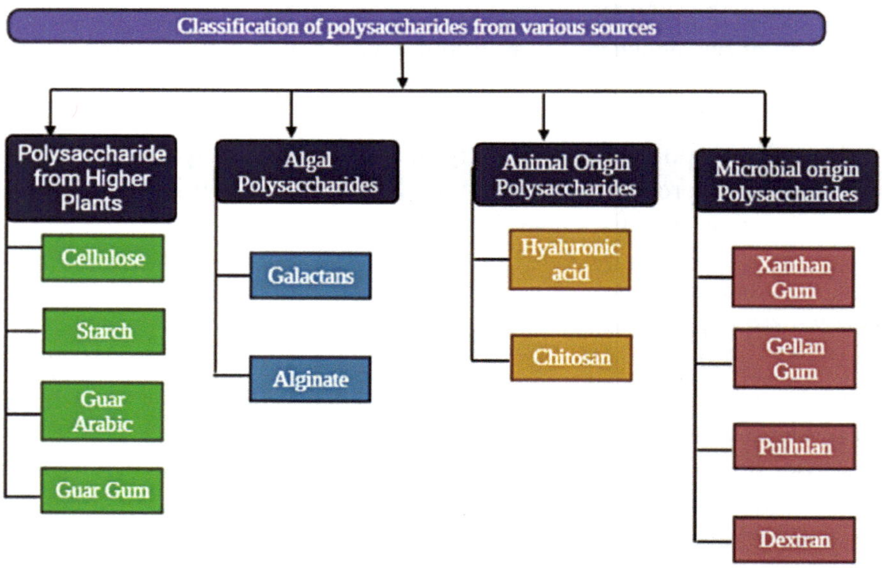

FIGURE 5.1 Classification of polysaccharides; describes the type of polysaccharides obtained from the different origins.

5.3.1 POLYSACCHARIDES FROM HIGHER PLANTS

The target cells must get these therapeutic genes in an effective and targeted manner. Serum nucleases frequently target pDNA, oligonucleotides, and short interfering RNA molecules as naked therapeutic gene targets. Lacklustre cellular uptake and poor therapeutic gene transfection efficiency are two issues with gene therapy. More than 1340 clinical studies using gene therapy have completed or are currently being done since the first one was conducted in 1989. The main barrier to human gene therapy's development is the lack of safe, effective, and regulated methods for delivering genes. An appropriate vector that can protect the genes and facilitate their dispersion to the target cells is therefore required for the ideal gene delivery system. It is believed that several polysaccharides based on nanoparticles are viable non-viral vectors for gene delivery systems. The DNA must be protected until it reaches its intended location in the ideal non-viral DNA delivery vector. Additionally, the gene's vector needs to be small enough for cells to easily absorb it into their cytoplasm. The ability of cationic carriers to form complexes with negatively charged pDNA and to condense and transfer DNA both *in vitro* and *in vivo* is how they are commonly characterized. Cationic polymers and cationic phospholipids are two leading candidates being researched as non-viral vectors for gene delivery. Another benefit of cationic lipids is that they are less stable than cationic polymers (Figure 5.1).

In recent years, a diverse polysaccharide utilized as synthetic cationic polymers. Non-viral vectors are becoming increasingly interested in chitosan derivatives as possible gene delivery solutions. Compress genes into nanoscale complexes, due to the principal amine groups in chitosan on the polymer's backbone, chitosan exhibits cationic charges that combine with anionic pDNA to form electrolyte complexes. Modified polysaccharide nanoparticles have shown to extend pDNA's blood circulation and improve transfection effectiveness. For instance, the pDNA was successfully transported by chitosan-graft-spermine and hydrophobized dextran-spermine copolymers both *in vitro* and *in vivo*. To be more exact, spermine and periodate-oxidized chitosan were combined to generate chitosan-graft-spermine (CHI-gSPE).

5.3.1.1 STARCH

Frequently occurring changes to starch and SNPs Starch has limited use in industry due to its weak solubility and high viscosity. Starch (Figure 5.2) can be chemically altered to enhance specific characteristics and add extra functional groups. The chain's many hydroxyl groups on the glucose units serve as locations for multifunctional group modification. So, different chemical modifications of starch have been researched. Anionic, cationic, and hydrophobic alterations are three popular types. By converting the hydroxyl groups in starch into carboxyl groups, one can produce anionic starch [24]. Hypochlorite oxidation and ozonation oxidation are often used to carry out the oxidation process. As a conventional oxidation agent, hypochlorite salt is simple to work within a commercial setting. The apparent disadvantage of hypochlorite oxidation is that it results in significant volumes of salt production that must be disposed of. Ozone, on the other hand, is a potent oxidizing agent that is clean and leaves no traces behind. However, the risk of lung disease from ozone exposure restricts the use of the gas in industry and research [25]. The addition of anionic groups to starch, such as carboxymethyl starch, is another method of modification. The presence of carboxymethyl functional groups enhances the solubility and lowers the temperature at which starch begins to gelatinize. To enhance the quality of food, carboxymethyl starch is frequently employed as a thickening, stabilizing, and water-retaining agent. By adding cationic groups to the backbone, cationic starch is frequently created. Through an etherification reaction with the hydroxyl groups of starch, cationic reagents with amino, ammonium, sulfonium, or phosphonium groups may be added [26, 27]. Traditionally, quaternary ammonium groups are used to etherify starch to create cationic starch. Cationic reagents can introduce ammonium groups, such as 3-chloro-2-hydroxypropyltrialkylammonium chloride and glycidyl trimethyl ammonium chloride. Cationic SNPs have recently been shown to be effective gene delivery systems. One of the most efficient ways to create hydrophobically modified starch with various levels of substitution is chemical grafting. The esterification of starch with long-chain fatty acids to create hydrophobic starch material has been shown in several articles. Starch acylation has proven to be the easiest and most economically successful hydrophobic alteration to date. In comparison to native starch, acylated starch is more thermally stable and has a lower gelatinization temperature. The exceptional qualities of acetylated starch have prospective uses in the packaging sector, as well as in surface coating and biomedical materials [28–32].

5.3.1.2 CELLULOSE

HEC (Figure 5.3) is highly soluble in water; it can be synthesized without the use of organic solvents. Structurally, like a PEG repeating unit, which is frequently used to increase serum stability in biomedical applications, is the side chain unit of HEC. It is logical to anticipate that HEC, which serves as the main component of gene/drug delivery systems (DDSs), can have excellent serum stability [33, 34]. Himmelein et al. [35] established an HEC-based

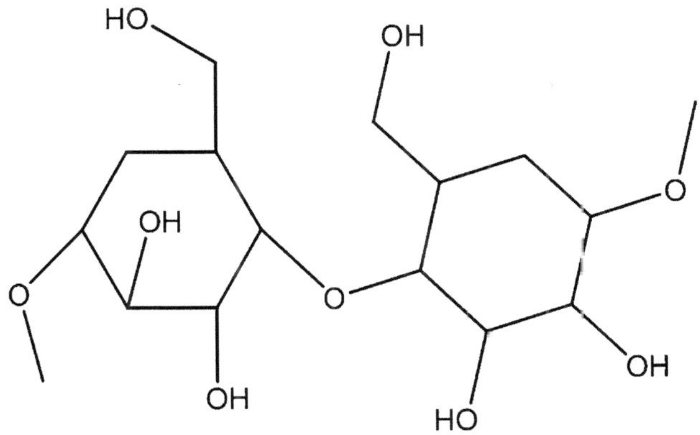

FIGURE 5.2 Simple starch.

hydrogel as a multiple 3D cross-linker, and an adamantane side group functionalized cellulose polymer was created. Additionally, the gel network revealed attentiveness since the presence of a rival host and guest may easily demolish it. Rheological analysis was used to study the material qualities, and it was discovered that the component concentrations could be easily adjusted. Most impressively, the hydrogel demonstrated shear-thinning and self-healing abilities, was injectable, and quickly recovered its bulk structure. The existence of vesicles inside the gel network and the components' biocompatibility may offer additional benefits for drug delivery. Ho et al. [36] increased the antibacterial activity of MTZ, by developing a gel made of HEC and MTZ

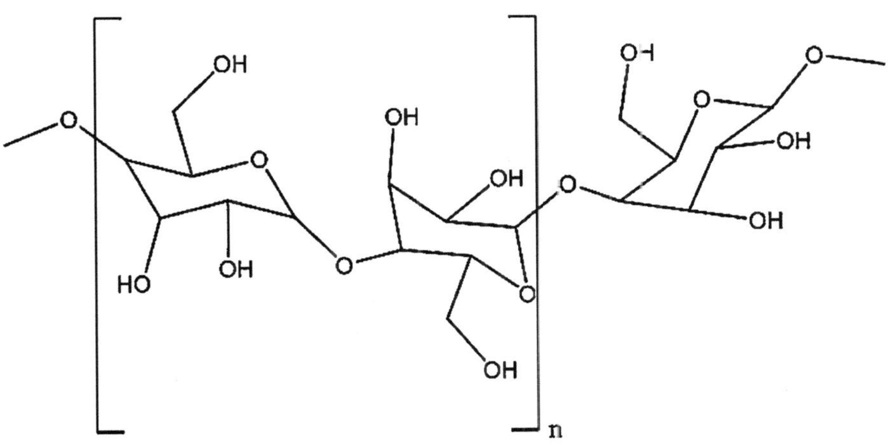

FIGURE 5.3 Structure of cellulose.

loaded in solid lipid nanoparticles. A combination of heat homogenization and solvent evaporation was used to create SLNs. According to the data, the synthetic SLNs had a narrow distribution and were around 200 nm in size. The HEC (3%, w/w)-based gel developed a uniform, smooth structure and exhibited favorable rheological and mechanical characteristics [36, 37].

5.3.1.3 GUAR GUM

Guar Gum (GG) (Figure 5.4) is a nontoxic, biocompatible, biodegradable, and inexpensive biopolymer generated from natural sources Because GGs are a flexible polymer with advantageous characteristics, including biocompatibility and biodegradability. It has a variety of characteristics, including physicochemical stability, GG is utilized as a thickening, stabilizing, suspending, and emulsifying ingredient in pharmaceutical technology. It is utilized as a binding and disintegrating agent in the creation of several solid medication administration dosage methods. As colonic fluids are promising as drug carriers, it has efficient gelling properties and enzymatic degradation. Chilly water causes GG to swell and create colloidal dispersions. The release of the medication from the dosage forms is delayed by the gelling network of GG. Therefore, physical and chemical functionalizations of GG have been used mostly for orally colon-targeted DDSs to manage its broad swelling features in different buffer solutions. In recent literature, a number of GG-based systems have been proposed as viable biomaterial carriers for regulated drug delivery.

For the delivery of ketoprofen (an NSAID used as a hydrophobic model medication), Prabaharan et al. [32] used microwave irradiation to create the GG-g-PCL copolymer. With longer microwave treatment times, the polycaprolactone grafting % rose. Due to its amphiphilic nature, it is assembled into micelles in an aqueous solution. According to a DLS analysis, the GG-g-PCL micelles had a restricted distribution with a diameter between 75 and 135 nm. The TEM was used to examine the spherical-shaped micelles, which had diameters between 60 and 100 nm. Ketoprofen, an anti-inflammatory medication, had a loading capacity of roughly 18.3 weight percent and was integrated into the micelles. The findings of research on the release of ketoprofen suggest that this approach seems to be a highly promising means of delivering hydrophobic medications through controlled release. The GG-g-PCL co-polymers *in vitro* degradation data demonstrate that breakdown is gradual for the first two weeks, which makes it a good drug carrier.

A thorough breakdown occurs after two weeks, paving the route for renal excretion. The GG-g-PCL micelles might undoubtedly be employed as drug carriers to enable regulated drug release, as has been shown throughout [32].

GG is a naturally occurring water-soluble and (2.5 106 g/mol) high molecular weight polysaccharide that undergoes hydrogen bonding in the aqueous solution due to hydroxyl groups. Inter-molecular chain tangling in water gives GG its gelling and thickening capabilities, as well as the ability to produce high viscosities even at low concentrations [38]. GG has the power to influence most systems through its behavior. The physicochemical and electrokinetic features of the systems can be changed by minute amounts of GG [39, 40].

GG lengthens the interaction time in the mucus lining of the tissue by interacting strongly with it. Due to its propensity to gel, enzymatic breakdown, and general regulatory acceptance, GG has been identified as a potential gene medication delivery vehicle due to its high hydrophilic character and invasive swelling matrix system-controlled drug delivery [41].

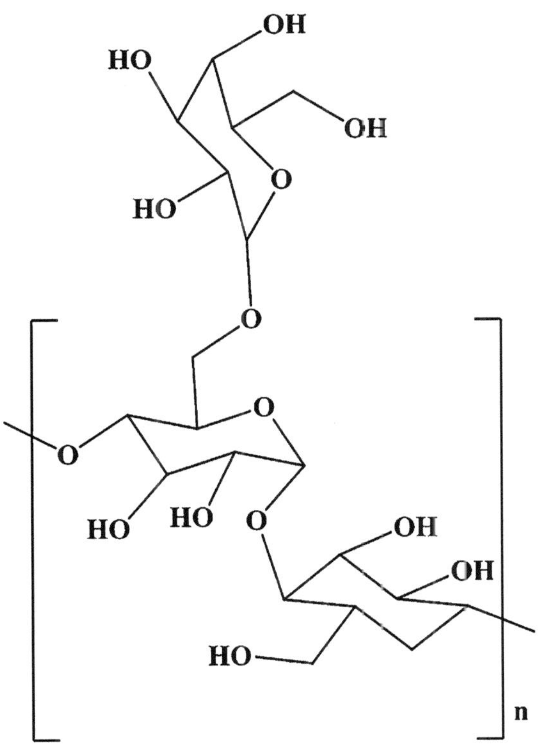

FIGURE 5.4 Structure of guar gum.

Seeli et al. [42] made barium cross-linked hybrid guar gum succinate (GGS) based beads and SA. GGS is a product of the reaction between guar gum and succinic anhydride. GGS may be employed as an excellent option due to its hydrophilic nature, which is suitable for colon-specific drug delivery. SA is a polysaccharide that dissolves in water and is made up of blocks of alternating 1–4 connected L-guluronic and D-mannuronic acid residues. It has potential for use in pharmacological and biological applications since it is a mucoadhesive, biodegradable, pH-sensitive, and biocompatible polymer. Due to the inclusion of SA and GGS, these beads may have enhanced pH sensitivity, better water absorption capacity, and improved drug release qualities.

GG is usually grafted with polyethyleneimine (PEI), a non-viral vector with a high buffering capacity, and is the industry standard for gene transfection [42]. Thus, LMW b-PEI has been grafted with PEG, other polysaccharides, and polyester to lessen its cytotoxicity and increase transfection. Chemically functionalized chitosan derivatives have been thoroughly investigated for gene delivery in polysaccharides, and GG has recently attracted attention. GG was originally used to transport a gene by grafting LMW b-PEI with CMGG, which has improved plasmid DNA (pDNA) binding properties [43, 44]. Create GNPs, GG was coupled to LPEI using a coupling agent based on naphthalic anhydride. The carrier has been discovered to generate spherical cytocompatibility complexes with exceptional gene binding abilities. Studies conducted *in vitro* revealed that the carrier's excellent buffering capacity contributed to its high transfection efficiency [44].

5.3.2 ALGAL POLYSACCHARIDES

5.3.2.1 ALGINATES

Another naturally occurring hydrophilic polymer is alginate, which is often produced from soil bacteria and sea-brown algae (e.g., *Laminaria hyperborean*). Alginate polymers are made up of blocks that often affect their physical characteristics. Alginates are mostly employed in the food and pharmaceutical sectors because of their appealing properties, such as gelation, non-toxicity, and water solubility [45]. Particularly when divalent cations like calcium and barium are present, external or internal gelation procedures for alginate are easily accomplished. The processes of swelling and erosion, which rely on the technique employed to create the gel, are primarily responsible for the sustained release of medications from alginate gel. Alginate gels, however, shrink in acidic environments but quickly degrade in alkaline environments. Alginate formulations are preferred for intestinal medication administration because of their pH-dependent swelling tendency [46]. Alginate (Figure 5.5) also has a special mucoadhesive characteristic that makes it helpful for entrapping different bioactive molecules, including cells, in synthetic extracellular matrix materials. Alginate hydrocolloid formulations provide the prolonged release of medicinal components [47]. Despite being beneficial for the local and long-term administration of several growth factors, alginate hydrogels' therapeutic uses have lagged behind those of other polysaccharides [48].

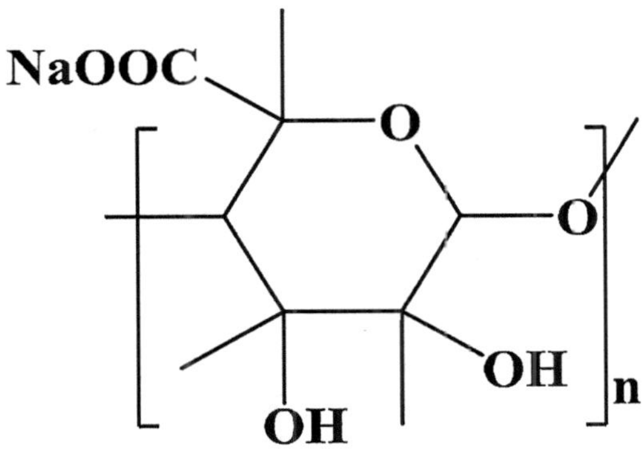

FIGURE 5.5 Structure of sodium alginate.

Proteins and peptides provide a variety of intrinsic difficulties when encapsulation and distribution. These include, but are not restricted to, the molecules' structural unfolding, aggregation, inherent instability, and instability. Sun et al. [49] employed a different method for the regulated distribution of proteins in this respect used solvent evaporation with poly (D,L-lactic-co-glycolic acid), and then adsorbed human serum albumin (HAS). After that, a calcium alginate hydrogel made from SA was used to plug the micropores in these porous biodegradable microspheres. With this type of protein lading, harmful chemicals, difficult dispensation conditions, and potential shear acquaintance that might result in protein denaturation avoided [50], HSA adsorption was found to be more effective at pH 4.5 as opposed to pH 7.4 phosphate buffer saline. At pH 4.5, the electrostatic interaction between HSA and PLGA predominated. Microsphere diameter very slightly increased as calcium alginate concentration was raised (from 0% to 2%), demonstrating that alginate was present in the porous network structure rather than the surface of the microspheres. Studies on microsphere deterioration in PBS show that compared to porous microspheres that weren't blocked, alginate blockage gave some degree of resistance to degradation by preventing water entry and distortion [51].

5.3.2.2 CHITOSAN

Chitosan (Figure 5.6) and its derivatives widely considered as drug and nucleic acid delivery carriers because to its global biocompatibility, low immunogenicity, biodegradability, and availability. It's noteworthy to note that the only cationic polysaccharide found naturally is chitosan. Chitosan is created via the b-glycosidic bonding [52–54]. Variable chitosans with different molecular may be created with variable amounts of deacetylation since chitin is normally alkaline-deacetylated to make it. Therefore, depending on the technique of manufacturing, the physicochemical and biological characteristics of chitosan products may change [55, 56]. Chemical modifications usually give chitosan its amphiphilicity, which leads to the emergence of matrix materials and self-assembling nanoparticles. Particularly, altered chitosan has shown success in delivering anti-cancer medications. Chitosan has also been widely employed in the transport of genes in plasmid DNA and siRNA also has immense importance to tissue engineering due to its cationic nature, which provides unique techniques for regrowing tissues to replace damaged organs. Proteoglycans and Glycosaminoglycans (GAGs) interact with cationic chitosan via electrostatic processes, pH sensitive. Chitosan-GAG interaction concentrates growth factors promoting tissue regeneration. They must possess biocompatibility and be non-toxic. Therefore, chitosan has undergone extensive research as a possible scaffold material for liver, bone, and articular cartilage tissue engineering [57, 58].

The chitosan-based nanoparticle drug delivery method has tremendous therapeutic potential by fusing the adaptability of siRNA and plasmid with the CS potential. DNA, miRNA, or other molecules can bind with chitosan and its derivatives by nucleic acids to influence the immune response, and preventive vaccines can be created [59]. As a result, efforts to use chitosan-based systems have considerably grown. The targeted delivery of siRNA to a particular cell type is accomplished by grafting a ligand onto the surface of the CS NP. The grafting strategies also help to identify a receptor that the cell expresses. The targeting ligand is essential for boosting the target cells' selective uptake and minimizing unintended deposits.

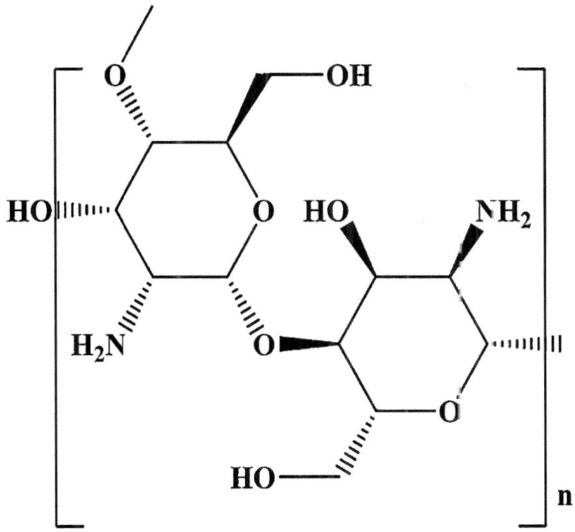

FIGURE 5.6 Structure of chitosan.

5.3.2.3 *HYALURONIC ACID*

All vertebrates have hyaluronan, a key component of synovial fluids, connective tissues, and vitreous humor. Hyaluronic acid (HA) is non-immunogenic, non-toxic, biocompatible, biodegradable [60]. The use of HA (Figure 5.7) or HA derivatives has been effective in a number of biological applications, including the treatment of arthritis, scaffolds for wound healing, tissue

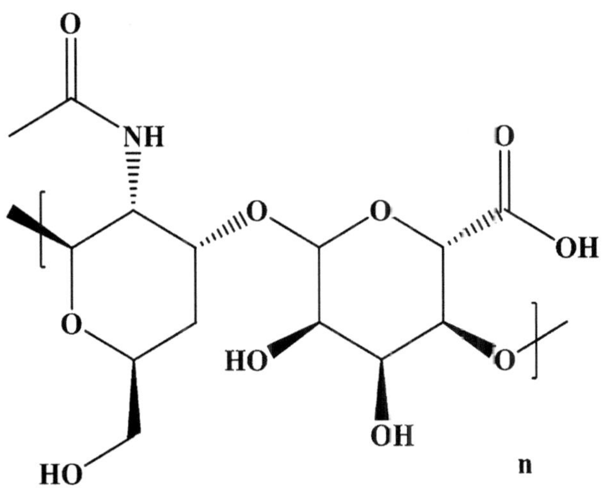

FIGURE 5.7 Structure of hyaluronic acid.

augmentation, eye surgery, and. Additionally, it has been well established that CD44 and CD168 interact with the HA receptor, and various tumor cells have been found to overexpress these cell surface glycoproteins. The absorption of HA nanoparticles by tumor cells can be considerably increased by this receptor-ligand interaction. As a result, many different HA nanoparticles have drawn interest for their potential application in the creation of a brand-new anti-cancer gene carrier [61].

This polymer's hyaluronan serves as a targeting agent in addition to protecting nanoparticles' surface charges [62, 64, 65]. Recent research has concentrated on creating hyaluronan integrated into the chitosan, subsequently forming complexes with other cationic polysaccharides, which also has the capacity to act as a bio adhesive. Hyaluronan-chitosan nanoparticles were created by Fuente et al. in 2008 [63] using a moderate ionotropic gelation process to transfer genetic material to the cornea and conjunctiva. Transfected with hyaluronan-chitosan nanoparticles to overcome certain barriers like blinking, tear flow, and penetration through epithelial tight junctions. In comparison with naked pEGFP, hyaluronan-chitosan polyplexes showed a higher expression of proteins. Interesting differences in gene expression were found between the two ocular cell lines, which may be explained by differences in mucin expression on their surfaces. Hyaluronan interacts with a variety of cellular receptors that are expressed on cell surfaces, making it a common choice for target-specific ligands.

Recently, A novel oral targeted vaccination carrier developed by Ya Liu uses nano-polyplex and oleoyl carboxymethyl-chitosan can store and concentrate growth factors helpful for tissue regeneration [66–70]. Hyaluronan is employed in the medical and pharmaceutical sectors as a non-viral vector as well as surgical assistance in ophthalmology. Hyaluronan may also improve the effectiveness of *in vivo* research due to its mucoadhesive qualities and selective interaction with hyaluronan-specific cell receptors.

5.3.3 POLYSACCHARIDES FROM MICROBIAL ORIGIN

5.3.3.1 CYCLODEXTRIN

Cyclodextrin (Figure 5.8) is a cyclic, cup-shaped, water-soluble polysaccharide produced by starch enzymatic breakdown [62]. One molecule of cyclodextrin has 21 hydroxyl groups, all of which may be used to create an infinite number of cyclodextrin derivatives [72, 73]. Dimethyl-cyclodextrin, hydroxypropyl-cyclodextrin, and methyl-cyclodextrin are just a few examples of the hydrophobic, hydrophilic, and ionic cyclodextrin derivatives have ability to increase the chemical stability, bioavailability, and solubility of drugs as well as decrease skin and eye irritation.

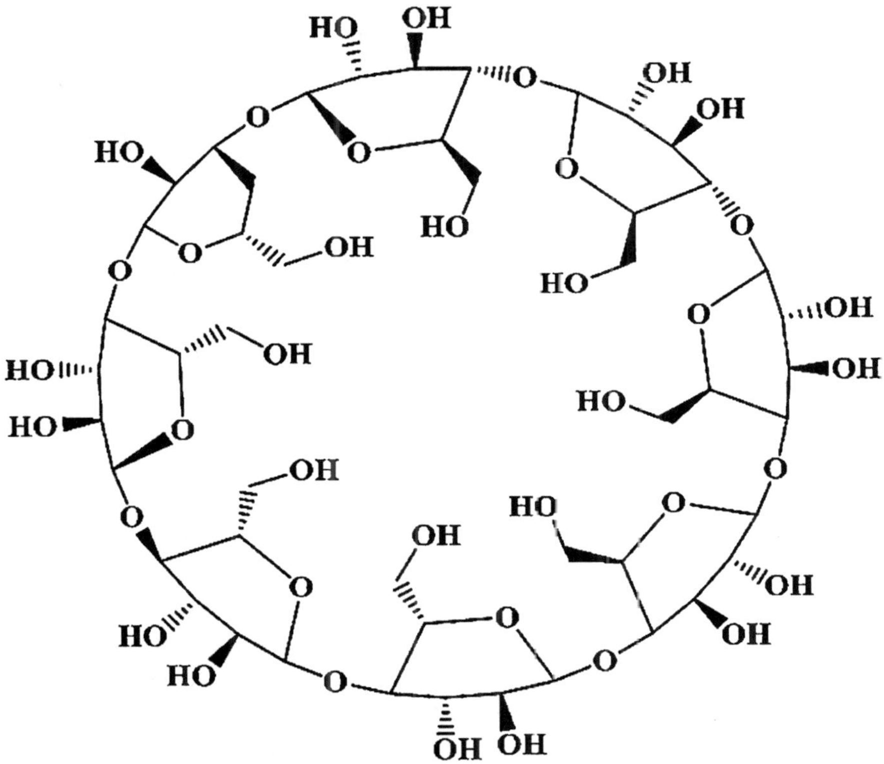

FIGURE 5.8 Structure of cyclodextrin.

Various modifications have been proposed to enhance the molecular selectivity and binding capacity of native cyclodextrins to overcome their inherent drawbacks. One such modification is the development of bridging bis(cyclodextrin)s. Cyclodextrins feature a hydrophobic inner cavity and a hydrophilic outer surface, promoting water solubility [74]. This unique structure allows cyclodextrins (hosts) to form inclusion complexes with small hydrophobic molecules (guests). In supramolecular chemistry, genetic materials can act as guest molecules, while cyclodextrins serve as host molecules [75, 76]. Li et al. [78] enhanced cyclodextrin-core nanoparticles by coating them with PEG, which reduced the polymer's cytotoxicity and increased p53 transfection efficiency by protecting the complex during delivery [77]. These polyplexes reduced tumor growth and tumor weight by 50%, highlighting the efficacy of the PEG coating. In another study, Li et al. [79] developed octa-arginine conjugated cyclodextrin with azobenzene-modified dextran for efficient gene delivery. Similarly, Jiang et al. [80] created a photosensitive non-viral vector using poly(-cyclodextrin) with azobenzene-terminated, cationic, linear, and branched DMAEMA [79]. Recent studies have described new cell-specific targeting agents utilizing cyclodextrin-modified polymers. These polymers, with cyclodextrins at their core, allow for multivalent attachment to various ligands. For example, Yang et al. [81] developed an oligoethylenimine (OEI)-cyclodextrin complex that showed reduced cytotoxicity while maintaining comparable gene transfection efficiency to branched PEI 25kDa. Grafting PEI onto the cyclodextrin backbone also enhanced buffering capacity and nucleic acid binding kinetics, without increasing cytotoxicity. Lai et al. [82] demonstrated that propane-1,2,3-triol-containing PEI-grafted cyclodextrin-DNA complexes had significantly improved DNA condensation and pH buffering properties, achieving efficient delivery with and without serum (10% fetal bovine serum). Cyclodextrin cross-linked with PEA has also been used to create star-shaped nanocomplexes with low cytotoxicity and high gene transport efficiency. Cyclodextrins have recently been modified with PEG, which improves water solubility, reduces aggregation, enhances serum stability, and slows opsonization during circulation. O'Mahony et al. [83] produced stable PEGylated cyclodextrin nanoparticles that reduced aggregation. Co-formulated cyclodextrin complexes exhibited lower absorption of fluorescently labeled siRNA in neuronal cells compared to cationic cyclodextrin [84]. Despite these advances, further modification is needed before PEGylated cyclodextrins can be effectively used for systemic distribution.

5.3.3.2 PULLULAN

Pullulan (Figure 5.9) is a repeating unit of (1–6)-linked maltotriosyl group a non-ionic, linear, water-soluble polysaccharide. Pullulan's distinctive linking pattern directs its physicochemical and biological characteristics, including Viscosity and adhesiveness [85, 86]. Due to its cationic and non-ionic nature modifications of pullulan can encapsulate pDNA resulted in nucleic acid delivery. In general, pullulan-modified nanoparticles especially helpful for liver targeting since they have a high affinity for glycoprotein receptors. Pullulan exhibits a superb adhesive quality as well, making it suitable for use in the creation of artificial extra-cellular matrix [87]. The use of pullulan-spermine/DNA anioplexes for gene transport to both primary cultured sensory neurones and explanted sensory ganglia helps studies of neuropathic pain and nerve regeneration. There are ramifications beyond *in vitro* neuronal applications for the synthesized from naturally occurring ingredients currently authorized for human use. *In-vitro* DNA delivery and considerable gene expression may be achieved using DEAE-pullulan matrices since they are nontoxic and possess the necessary chemical capabilities. This strategy has the potential to inform the development of novel implantable biomedical devices for *in vivo* gene therapy.

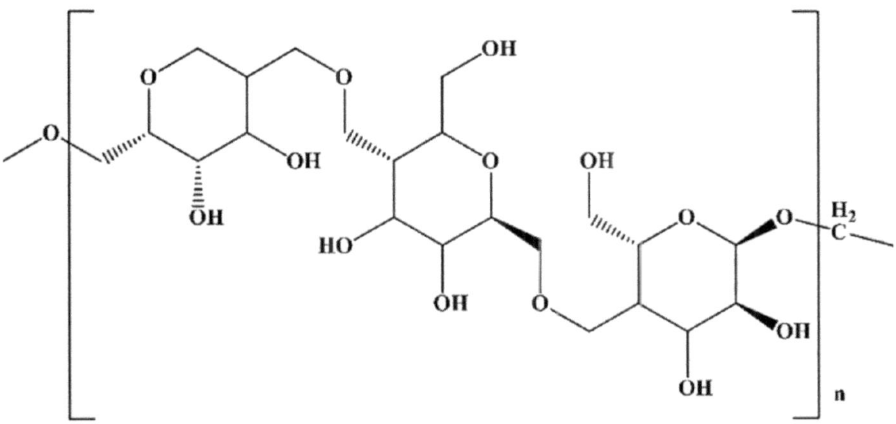

FIGURE 5.9 Structure of pullulan.

5.3.3.3 *XANTHAN GUM*

The gram-negative, aerobic pathogen Xanthomonas campestris produces the extracellular hetero polysaccharide known as xanthan gum (XG) (Figure 5.10) through fermentation [88–90]. It has been created to assemble HA with other biocompatible polymers to modify the mechanical and rheological characteristics of HA. Additionally, a mixture of HA and its advantages might be fully utilized by biocompatible polymers to make up for several shortcomings of other synthetic or biopolymers.

Several hybrid hydrogels have been researched by conjugating HA with natural [47, 48] and synthetic polymers [49, 50] to delve further into hydrogels or nanogels for improved drug delivery. Because of their distinctive advantages (such as the simplicity of curbing the component and nanogel

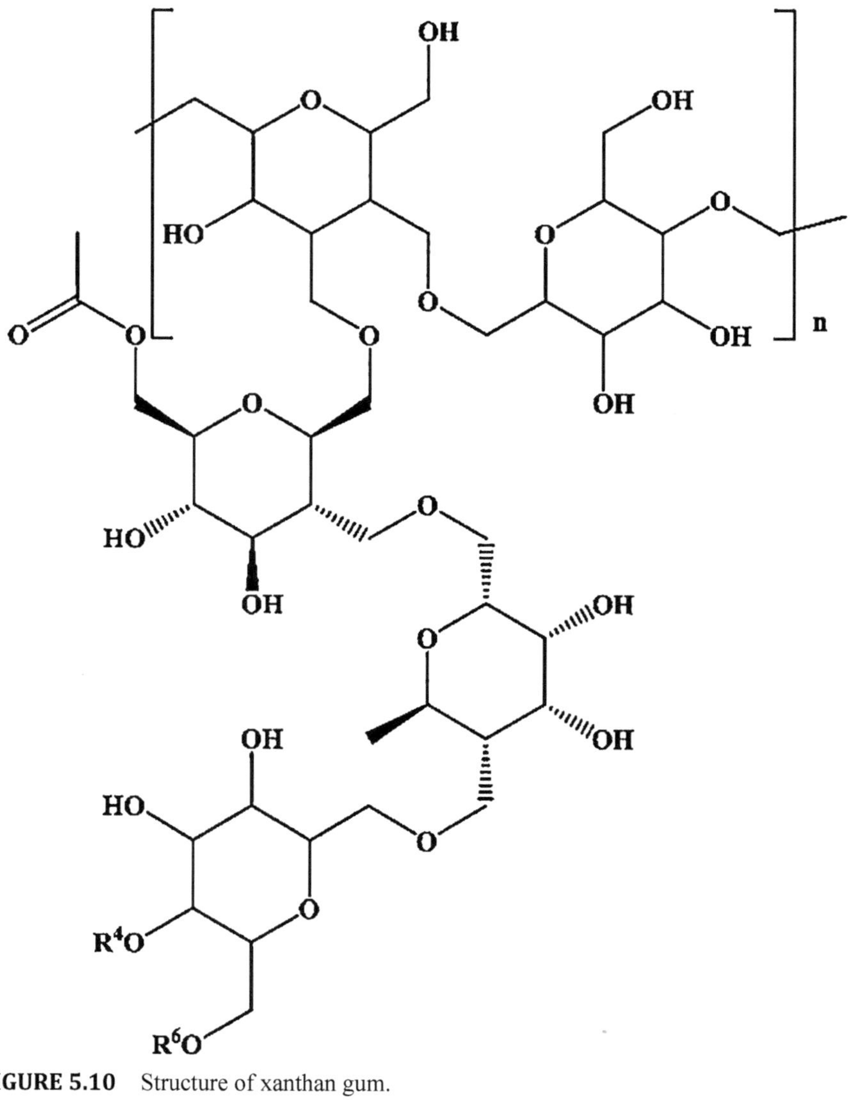

FIGURE 5.10 Structure of xanthan gum.

structure), synthetic polymers hold promise. HA microgel and methyl polyethylene glycol (mPEG) have shown remarkable biocompatibility. It is sought after to combine HA with anticancer medications to gain numerous advantages, such as greater solubility, prolonged plasma half-life, and better pharmacokinetic (PK) features. The signaling biomolecule HA is involved in leukocyte homing, cell adhesion, growth, and migration, among other biological processes. One notable advantage of using HA assembly in anticancer drug administration is passive tumor targeting. As a result, various efforts have produced a prolonged and amplificatory curative action.

XG has been widely utilized in the food industry as a thickening agent and stabilizer. Under the brand name Kelzan®, the Kelco firm (CP Kelco, Atlanta, GA, USA) began producing XG in the country in the 1960s. With the brand name Rhodopol® and Jungbunzlauer Austria AG in Austria, Solvay in Europe, and other companies began producing XG on a large scale. China has been one of the largest XG manufacturers since 2005. In the petroleum sector, it is added to water-based drilling fluids because of its pseudo-plastic behavior and heat stability. In the food and cosmetics sectors, XG primarily supports emulsion stability and serves as an effective thickener. XG is regarded as an active excipient in the pharmaceutical business for the prolonged release of medications. It is also employed as a drug delivery method because of its great stability across a wide variety of temperatures, pH levels, and ionic strengths, as well as its biocompatibility with the human body. Nanoparticles are also reduced and coated with XG to make them non-toxic and biodegradable.

According to the grafting theory, side chain oligomers are bound to hydrophilic polysaccharides with anionic surfaces of double-strand DNA strands to create three-dimensional contact. The structural characteristics of cationic polysaccharides determine their capacity to transfect cells and transport genes [88]. For instance, sorbitan monooleate nanoparticles were added to XG to target liver sinusoidal endothelial cells [92, 93]. After systemic injection, it was shown that XG span nanoparticles are absorbed by vascular endothelial cells lacking the mannose receptor in addition to LSECs [71]. The nanoparticles included the plasmid that produces more green fluorescent protein [91]. As therapeutic agents, protein- and peptide-based therapies provide a broad range of possibilities since they are more efficient and safer than chemical medications. Their usage has been limited, however, by delivery issues. Transdermal administration has been an alternative for delivery but has limited potential for absorption, and their oral bioavailability is exceedingly poor. For instance, xanthan gum complexed with poly (N-vinyl imidazole) was used to boost the efficiency of bovine serum albumin. A variety of pH condition studies were conducted since the carrier was pH sensitive. The non-Fickian release model showed that release occurred more rapidly at physiological pH 7.4 than at pH 1.2.

The loaded carrier was harmless to healthy cells. Staphylococcus aureus and *Aspergillus Niger* were both resistant to the antimicrobial effects of protein-loaded graft, demonstrating their value as antimicrobial drug transporters. PVI (poly-N-vinyl imidazole) complexes with xanthum gum hydrogel were a different carrier system designed for the delivery of BSA. Studies on cytotoxicity revealed that the hydrogel had acceptable biocompatibility. Additionally, SDS-PAGE was used to confirm the structure's intactness. The findings led to the conclusion that proteins and peptides might be delivered using XG-based DDSs.

5.4 CONCLUSION

Graft copolymerization is a successful way to add beneficial qualities to the primary polymer backbone, and these features have numerous uses in many industries. It is simple to add new qualities to a natural polymer using a synthetic polymer through grafting, with little loss of the substrate's original features. An effective method for adding the desired characteristic to polysaccharide is graft co-polymerization. Additionally, more stable than its virginal counterpart., grafted polysaccharide. Nucleic acids generally have serious drawbacks, such as poor stability, limited bioactivity, and unwanted side effects. Polysaccharides were discovered to offer the most promise among these polymers for gene delivery in biopharmaceutical and biomedical nanotechnology applications. To take advantage of their desirable qualities, hydrophobicity, biodegradability, non-toxicity, and cheap cost, polysaccharide-based nanoparticles are effective carriers; these polysaccharide nanoparticles may readily carry genes and enhance target-specificity.

Major developments in genomics, including the discovery of the structural and functional aspects of genes in the human genome and improvements in the management of plasmid-based gene expression systems, have lately increased the potential of grafting in gene therapy. Regarding the use of non-viral gene delivery vehicles for human therapy, there are still many issues that need to be resolved, and polysaccharides will be crucial in this. In conclusion, gene delivery is still a remarkably interesting field of study with a lot of promise for gene therapy in the future. *In vitro* gene delivery methods have undergone remarkable efforts to optimize, yet *in vivo* use may call for various vector properties. Numerous obstacles must be overcome for the grafting of polysaccharides in gene therapy to succeed, including biodistribution to first-pass organs, quick, complicated clearance, lack of tissue targeting, toxicity, nonspecific interactions, and so forth.

KEYWORDS

- **biopolymers**
- **copolymerization**
- **gene delivery**
- **grafting of polymers**
- **guar gum**
- **polysaccharides**

REFERENCES

1. Scheller, E. L., & Krebsbach, P. H. (2009). Gene therapy: Design and prospects for craniofacial regeneration. *Journal of Dental Research, 88*(7), 585–596.
2. Sarkar, T., Sarkar, S., & Gangopadhyay, D. N. (2020). Gene therapy and its application in dermatology. *Indian Journal of Dermatology, 65*(5), 341.
3. Kurrikoff, K., Gestin, M., & Langel, Ü. (2016). Recent *in vivo* advances in cell-penetrating peptide-assisted drug delivery. *Expert Opinion on Drug Delivery, 13*(3), 373–387.
4. Mulia, G. E., Picanço-Castro, V., Stavrou, E. F., Athanassiadou, A., & Figueiredo, M. L. (2021). Advances in the development and the applications of nonviral, episomal vectors for gene therapy. *Human Gene Therapy, 32*(19–20), 1076–1095.
5. Maeder, M. L., & Gersbach, C. A. (2016). Genome-editing technologies for gene and cell therapy. *Molecular Therapy, 24*(3), 430–446.
6. Cevher, E., Sezer, A. D., & Çağlar, E. (2012). Gene delivery systems: Recent progress in viral and non-viral therapy. In *Recent Advances in Novel Drug Carrier Systems* (pp. 437–470). Springer.
7. Bulcha, J. T., Wang, Y., Ma, H., Tai, P. W. L., & Gao, G. (2021). Viral vector platforms within the gene therapy landscape. *Signal Transduction and Targeted Therapy, 6*(1), 53.
8. Sung, Y. K., & Kim, S. W. (2019). Recent advances in the development of gene delivery systems. *Biomaterials Research, 23*, 1–7.
9. Jäger, T., Mokos, A., & Prasianakis, N. I. (2022). Pore-level multiphase simulations of realistic distillation membranes for water desalination. *Membranes, 12*(11), 1112.
10. Yenagolla, P., Sandeep, K., & Sharma, J. V. (2022). Natural polymers and their applications: A review. *EPRA International Journal of Research and Development (IJRD), 7*(1), 6–15.
11. Singh, P., Manzoor, K., & Ikram, S. (2021). *Innovation in Nano-Polysaccharides for Eco-Sustainability: From Science to Industrial Applications*. Elsevier.
12. Sivaramakrishnan, S., & Vimala, R. T. (2022). *Bacterial Bio Flocculant for Multifunctional Features*. Academic Press.
13. Rai, R., Alwani, S., & Badea, I. (2019). Polymeric nanoparticles in gene therapy: New avenues of design and optimization for delivery applications. *Polymers, 11*(4), 745.

14. Diekjürgen, D., & Grainger, D. W. (2017). Polysaccharide matrices used in 3D *in vitro* cell culture systems. *Biomaterials, 141*, 96–115.
15. Jin, M., Shi, J., Zhu, W., Yao, H., & Wang, D. A. (2021). Polysaccharide-based biomaterials in tissue engineering: A review. *Tissue Engineering Part B: Reviews, 27*(6), 604–626.
16. Hasnain, M. S., & Nayak, A. K. (2019). *Natural Polysaccharides in Drug Delivery and Biomedical Applications*. Academic Press.
17. Mohammadi, K., Sani, M. A., Azizi-Lalabadi, M., & McClements, D. J. (2022). Recent progress in the application of plant-based colloidal drug delivery systems in the pharmaceutical sciences. *Advances in Colloid and Interface Science, 302*, 102734.
18. Naveen, C., & Shastri, N. R. (2019). Polysaccharide nano micelles as drug carriers. In *Polysaccharide Carriers for Drug Delivery* (pp. 339–363). Woodhead Publishing.
19. Kumar, D., Pandey, J., Raj, V., & Kumar, P. (2017). A review on the modification of polysaccharide through graft copolymerization for various potential applications. *The Open Medicinal Chemistry Journal, 11*, 109–122.
20. Singh, B., Khurana, R. K., Garg, B., Saini, S., & Kaur, R. (2017). Stimuli-responsive systems with diverse drug delivery and biomedical applications: Recent updates and mechanistic pathways. *Critical Reviews™ in Therapeutic Drug Carrier Systems, 34*(3), 209–255.
21. Jha, S., Malviya, R., Fuloria, S., Sundram, S., Subramaniyan, V., Sekar, M., Sharma, P. K., Chakravarthi, S., Wu, Y. S., Mishra, N., & Meenakshi, D. U. (2022). Characterization of microwave-controlled polyacrylamide graft copolymer of tamarind seed polysaccharide. *Polymers, 14*(5), 1037. https://doi.org/10.3390/polym14051037.
22. Thakur, M. K., Thakur, V. K., Gupta, R. K., & Pappu, A. (2016). Synthesis and applications of biodegradable soy-based graft copolymers: A review. *ACS Sustainable Chemistry & Engineering, 4*(1), 1–7. https://doi.org/10.1021/acssuschemeng.5b00814.
23. Hada, D., Rathore, K., Barupal, T., Chundawat, N. S., Sharma, K., & Chauhan, N. P. (2019). Grafted biopolymers I: Methodology and factors affecting grafting. In *Advanced Functional Polymers for Biomedical Applications* (pp. 21–42). Elsevier.
24. Jäger, T., Mokos, A., Prasianakis, N. I., & Leyer, S. (2022). Pore-level multiphase simulations of realistic distillation membranes for water desalination. *Membranes, 12(11), 1112.*
25. Cuerda-Correa, E. M., Alexandre-Franco, M. F., & Fernández-González, C. (2019). Advanced oxidation processes for the removal of antibiotics from water: An overview. *Water, 12*(1), 102.
26. Chakraborty, I., Mal, S. S., Paul, U. C., Rahman, M. H., & Mazumder, N. (2022). An insight into the gelatinization properties influencing the modified starches used in the food industry: A review. *Food and Bioprocess Technology, 15*(6), 1195–1223.
27. Ojogbo, E., Ogunsona, E. O., & Mekonnen, T. H. (2020). Chemical and physical modifications of starch for renewable polymeric materials. *Materials Today Sustainability, 7*, 100028.
28. Sharma, M., Aguado, R., Murtinho, D., Valente, A. J., De Sousa, A. P., & Ferreira, P. J. (2020). A review on cationic starch and nanocellulose as paper coating components. *International Journal of Biological Macromolecules, 162*, 578–598.
29. Gopinath, V., Kamath, S. M., Priyadarshini, S., Chik, Z., Alarfaj, A. A., & Hirad, A. H. (2022). Multifunctional applications of natural polysaccharide starch and cellulose: An update on recent advances. *Biomedicine & Pharmacotherapy, 146*, 112492.

30. Falua, K. J., Pokharel, A., Babaei-Ghazvini, A., Ai, Y., & Acharya, B. (2022). Valorization of starch to biobased materials: A review. *Polymers, 14*(11), 2215.
31. Torres, F. G., & De-la-Torre, G. E. (2022). Synthesis, characteristics, and applications of modified starch nanoparticles: A review. *International Journal of Biological Macromolecules, 194*, 289–305.
32. Tiwari, A., & Prabaharan, M. (2010). An amphiphilic nanocarrier based on guar gum-graft-poly (ε-caprolactone) for potential drug-delivery applications. *Journal of Biomaterials Science, Polymer Edition, 21*(6–7), 937–949.
33. Gomez, I., Alesanco, Y., Blázquez, J. A., Viñuales, A., & Colmenares, L. C (2020). Room-temperature self-standing cellulose-based hydrogel electrolytes for electrochemical devices. *Polymers, 12*(11), 2686.
34. Kamel, S., Ali, N., Jahangir, K., Shah, S. M., & El-Gendy, A. A. (2008). Pharmaceutical significance of cellulose: A review. *Express Polymer Letters, 2*(11), 758–778.
35. Himmelein, S., Lewe, V., Stuart, M. C., & Ravoo, B. J. (2014). A carbohydrate-based hydrogel containing vesicles as responsive non-covalent cross-linkers. *Chemical Science, 5*(3), 1054–1058.
36. Ho, H. N., Le, H. H., Le, T. G., Duong, T. H., Ngo, V. Q., Dang, C. T., Tran, T. H., & Nguyen, C. N. (2022). Formulation and characterization of hydroxyethyl cellulose-based gel containing metronidazole-loaded solid lipid nanoparticles for buccal mucosal drug delivery. *International Journal of Biological Macromolecules, 194*, 1010–1018.
37. Fayazpour, F., Lucas, B., Alvarez-Lorenzo, C., Sanders, N. N., Demeester, J., & De Smedt, S. C. (2006). Physicochemical and transfection properties of cationic hydroxyethyl cellulose/DNA nanoparticles. *Biomacromolecules, 7*(10), 2856–2862.
38. Shoaib, M. H., Sikandar, M., Ahmed, F. R., Ali, F. R., Qazi, F., Yousuf, R. I., Irshad, A., Jabeen, S., & Ahmed, K. (2021). Applications of polysaccharides in controlled release drug delivery systems. In *Polysaccharides: Properties and Applications* (pp. 607–656). CRC Press.
39. George, A., Shah, P. A., & Shrivastav, P. S. (2019). Guar gum: Versatile natural polymer for drug delivery applications. *European Polymer Journal, 112*, 722–735.
40. Saya, L., Malik, V., Singh, A., Singh, S., Gambhir, G., Singh, W. R., Chandra, R., & Hooda, S. (2021). Guar gum-based nanocomposites: Role in water purification through efficient removal of dyes and metal ions. *Carbohydrate Polymers, 261*, 117851.
41. Ugoeze, K. C. (2020). Bioadhesive polymers for drug delivery applications. In *Bio Adhesives in Drug Delivery* (pp. 29–56). Elsevier.
42. Seeli, D. S., Dhivya, S., Selvamurugan, N., & Prabaharan, M. (2016). Guar gum succinate-sodium alginate beads as a pH-sensitive carrier for colon-specific drug delivery. *International Journal of Biological Macromolecules, 91*, 45–50.
43. Cheng, W., Yang, C., Hedrick, J. L., Williams, D. F., Yang, Y. Y., & Ashton-Rickardt, P. G. (2013). Delivery of a granzyme B inhibitor gene using carbamate-mannose modified PEI protects against cytotoxic lymphocyte killing. *Biomaterials, 34*(14), 3697–3705.
44. Yu, K., Wu, S., & Li, H. (2016). A chitosan-graft-PEI-eprosartan conjugate for cardiomyocyte-targeted VEGF plasmid delivery in myocardial ischemia gene therapy. *Journal of Experimental Nanoscience, 11*(2), 81–96.
45. Taranejoo, S., Chandrasekaran, R., Cheng, W., & Hourigan, K. (2016). Bioreducible PEI-functionalized glycol chitosan: A novel gene vector with reduced cytotoxicity and improved transfection efficiency. *Carbohydrate Polymers, 153*, 160–168.

46. Nesamony, J., Singh, P. R., Nada, S. E., Shah, Z. A., & Kolling, W. M. (2012). Calcium alginate nanoparticles synthesized through a novel interfacial cross-linking method as a potential protein drug delivery system. *Journal of Pharmaceutical Sciences, 101*(6), 2177–2184.
47. Zhao, J., Zhao, X., Guo, B., & Ma, P. X. (2014). Multifunctional interpenetrating polymer network hydrogels based on methacrylated alginate for the delivery of small molecule drugs and sustained release of protein. *Biomacromolecules, 15*(9), 3246–3252.
48. El-Sherbiny, I. M. (2010). Enhanced pH-responsive carrier system based on alginate and chemically modified carboxymethyl chitosan for oral delivery of protein drugs: Preparation and in-vitro assessment. *Carbohydrate Polymers, 80*(4), 1125–1136.
49. Sun, Y., Zhao, Y., Teng, S., Hao, F., Zhang, H., Meng, F., Zhao, X., Zheng, X., Bi, Y., Yao, Y., & Lee, R. J. (2019). Folic acid receptor-targeted human serum albumin nanoparticle formulation of cabazitaxel for tumor therapy. *International Journal of Nanomedicine, 14*, 135–148.
50. Bajwa, N., Singh, P. A., Jyoti, K., Baldi, A., Chandra, R., & Madan, J. (2021). Scale-up, preclinical, and clinical status of poly (lactide-co-glycolide) and its copolymers-based drug delivery systems. In *Nanomaterials: Evolution and Advancement Towards Therapeutic Drug Delivery (Part II)* (pp. 246–262). Springer.
51. Xie, C., Xiong, Q., Wei, Y., Li, X., Hu, J., He, M., Wei, S., Yu, J., Cheng, S., Ahmad, M., & Liu, Y. (2023). Fabrication of biodegradable hollow microsphere composites made of polybutylene adipate co-terephthalate/polyvinylpyrrolidone for drug delivery and sustained release. *Materials Today Bio, 20*, 100628.
52. Koshy, S. T., Zhang, D. K., Grolman, J. M., Stafford, A. G., & Mooney, D. J. (2018). Injectable nanocomposite cryogels for versatile protein drug delivery. *Acta Biomaterialia, 65*, 36–43.
53. Nunthanid, J., Laungtana-Anan, M., Sriamornsak, P., Limmatvapirat, S., Puttipipatkhachorn, S., Lim, L. Y., & Khor, E. (2004). Characterization of chitosan acetate as a binder for sustained release tablets. *Journal of Controlled Release, 99*(1), 15–26.
54. Chae, S. Y., Son, S., Lee, M., Jang, M. K., & Nah, J. W. (2005). Deoxycholic acid-conjugated chitosan oligosaccharide nanoparticles for efficient gene carrier. *Journal of Controlled Release, 109*(1–3), 330–344.
55. Gao, Y., Zhang, Z., Chen, L., Gu, W., & Li, Y. (2009). Chitosan N-betainates/DNA self-assembly nanoparticles for gene delivery: *In vitro* uptake and transfection efficiency. *International Journal of Pharmaceutics, 371*(1–2), 156–162.
56. Abourehab, M. A., Pramanik, S., Abdelgawad, M. A., Abualsoud, B. M., Kadi, A., Ansari, M. J., & Deepak, A. (2022). Recent advances of chitosan formulations in biomedical applications. *International Journal of Molecular Sciences, 23*(18), 10975.
57. Wasupalli, G. K., & Verma, D. (2022). Thermosensitive injectable hydrogel based on chitosan-polygalacturonic acid polyelectrolyte complexes for bone tissue engineering. *Carbohydrate Polymers, 294*, 119769.
58. Dash, M., Chiellini, F., Ottenbrite, R. M., & Chiellini, E. (2011). Chitosan—A versatile semi-synthetic polymer in biomedical applications. *Progress in Polymer Science, 36*(8), 981–1014.
59. Hari, S. K., Gauba, A., Shrivastava, N., Tripathi, R. M., Jain, S. K., & Pandey, A. K. (2023). Polymeric micelles and cancer therapy: An ingenious multimodal tumor-targeted drug delivery system. *Drug Delivery and Translational Research, 13*(1), 135–163.

60. Cheng, Y. (2018). *Polymeric Gene Delivery Systems*. Springer.
61. Abatangelo, G., Vindigni, V., Avruscio, G., Pandis, L., & Brun, P. (2020). Hyaluronic acid: Redefining its role. *Cells, 9*(7), 1743.
62. Hong, S. J., Ahn, M. H., Sangshetti, J., Choung, P. H., & Arote, R. B. (2018). Sugar-based gene delivery systems: Current knowledge and new perspectives. *Carbohydrate Polymers, 181*, 1180–1193.
63. De la Fuente, M., Seijo, B., & Alonso, M. J. (2008). Novel hyaluronic acid-chitosan nanoparticles for ocular gene therapy. *Investigative Ophthalmology & Visual Science, 49*(5), 2016–2024.
64. Bachu, R. D., Chowdhury, P., Al-Saedi, Z. H., Karla, P. K., & Boddu, S. H. (2018). Ocular drug delivery barriers—Role of nanocarriers in the treatment of anterior segment ocular diseases. *Pharmaceutics, 10*(1), 28.
65. Aibani, N., Rai, R., Patel, P., Cuddihy, G., & Wasan, E. K. (2021). Chitosan nanoparticles at the biological interface: Implications for drug delivery. *Pharmaceutics, 13*(10), 1686.
66. Liu, Y., Wang, F. Q., Shah, Z., Cheng, X. J., Kong, M., Feng, C., & Chen, X. G. (2016). Nano-polyplex based on oleoyl-carboxymethy-chitosan (OCMCS) and hyaluronic acid for oral gene vaccine delivery. *Colloids and Surfaces B: Bio Interfaces, 145*, 492–501.
67. Jurczyk, M., Jelonek, K., Musiał-Kulik, M., Beberok, A., Wrześniok, D., & Kasperczyk, J. (2021). Single-versus dual-targeted nanoparticles with folic acid and biotin for anticancer drug delivery. *Pharmaceutics, 13*(3), 326.
68. Zeng, Y., Xiang, Y., Sheng, R., Tomás, H., Rodrigues, J., Gu, Z., Zhang, H., Gong, Q., & Luo, K. (2021). Polysaccharide-based nanomedicines for cancer immunotherapy: A review. *Bioactive Materials, 6*(10), 3358–3382.
69. Srinivasan, B., Kolli, A. R., Esch, M. B., Abaci, H. E., Shuler, M. L., & Hickman, J. J. (2015). TEER measurement techniques for *in vitro* barrier model systems. *Journal of Laboratory Automation, 20*(2), 107–126.
70. Guter, M., & Breunig, M. (2017). Hyaluronan is a promising excipient for ocular drug delivery. *European Journal of Pharmaceutics and Biopharmaceutics, 113*, 34–49.
71. Cramer, S., Rempe, R., & Galla, H. J. (2012). Exploiting the properties of biomolecules for brain targeting of nanoparticulate systems. *Current Medicinal Chemistry, 19*(19), 3163–3187.
72. Pedersen, C. M., & Bols, M. (2013). Cyclodextrin-based artificial enzymes: Synthesis and function. *Organic Synthesis and Molecular Engineering*, 305–332.
73. Omtvedt, L. A., Kristiansen, K. A., Strand, W. I., Aachmann, F. L., Strand, B. L., & Zaytseva-Zotova, D. S. (2021). Alginate hydrogels functionalized with β-cyclodextrin as a local paclitaxel delivery system. *Journal of Biomedical Materials Research Part A, 109*(12), 2625–2639.
74. Rajalakshmi, P., Peter, D. N., & Ananthi, N. (2021). Structure-activity relationship of supramolecular compounds in drug delivery. *Mini-Reviews in Organic Chemistry, 18*(7), 961–989.
75. Wankar, J., Kotla, N. G., Gera, S., Rasala, S., Pandit, A., & Rochev, Y. A. (2020). Recent advances in host-guest self-assembled cyclodextrin carriers: Implications for responsive drug delivery and biomedical engineering. *Advanced Functional Materials, 30*(44), 1909049.
76. Periasamy, R. (2020). A systematic review on the significant roles of cyclodextrins in the construction of supramolecular systems and their potential usage in various fields. *Journal of Carbohydrate Chemistry, 39*(5–6), 189–216.

77. Hu, Y., Yuan, W., Zhao, N. N., Ma, J., Yang, W. T., & Xu, F. J. (2013). Supramolecular pseudo-block gene carriers based on bioreducible star polycations. *Biomaterials, 34*(21), 5411–5422.
78. Li, W., Du, J., Zheng, K., Zhang, P., Hu, Q., & Wang, Y. (2014). Multifunctional nanoparticles via host-guest interactions: A universal platform for targeted imaging and light-regulated gene delivery. *Chemical Communications, 50*(13), 1579–1581.
79. Li, W., Liu, Y., Du, J., Ren, K., & Wang, Y. (2015). Cell penetrating peptide-based polyplexes shelled with polysaccharide to improve stability and gene transfection. *Nanoscale, 7*(18), 8476–8484.
80. Jiang, Q., Zhang, Y., Zhuo, R., & Jiang, X. (2016). Supramolecular host-guest polycationic gene delivery system based on poly (cyclodextrin) and azobenzene-terminated polycations. *Colloids and Surfaces B: Bio Interfaces, 147*, 25–35.
81. Yang, C., Li, H., Goh, S. H., & Li, J. (2007). Cationic star polymers consisting of α-cyclodextrin core and oligoethylenimine arms as nonviral gene delivery vectors. *Biomaterials, 28*(21), 3245–3254.
82. Lai, W. F., & Jung, H. S. (2014). Cell transfection with a β-cyclodextrin-PEI-propane-1, 2, 3-triol nanopolymer. *PLOS ONE, 9*(6), e100258.
83. O'Mahony, A. M., Ogier, J., Darcy, R., Cryan, J. F., & O'Driscoll, C. M. (2013). Cationic and PEGylated amphiphilic cyclodextrins: Co-formulation opportunities for neuronal siRNA delivery. *PLOS ONE, 8*(6), e66413.
84. Godinho, B. M., Ogier, J. R., Quinlan, A., Darcy, R., Griffin, B. T., Cryan, J. F., & O'Driscoll, C. M. (2014). PEGylated cyclodextrins as novel siRNA nanosystems: Correlations between polyethylene glycol length and nanoparticle stability. *International Journal of Pharmaceutics, 473*(1–2), 105–112.
85. Sharma, V. K., Liu, X., Oyarzún, D. A., Abdel-Azeem, A. M., Atanasov, A. G., Hesham, A. E., Barik, S. K., Gupta, V. K., & Singh, B. N. (2022). Microbial polysaccharides: An emerging family of natural biomaterials for cancer therapy and diagnostics. In *Seminars in Cancer Biology* (Vol. 86, pp. 706–731). Academic Press.
86. Tabasum, S., Noreen, A., Maqsood, M. F., Umar, H., Akram, N., Chatha, S. A., & Zia, K. M. (2021). A review on versatile applications of blends and composites of pullulan with natural and synthetic polymers. *International Journal of Biological Macromolecules, 120*(A), 603–632
87. Ganie, S. A., Rather, L. J., & Li, Q. (2021). A review on anticancer applications of pullulan and pullulan derivative nanoparticles. *Carbohydrate Polymer Technologies and Applications, 2*, 100115.
88. Jadav, M., Pooja, D., Adams, D. J., & Kulhari, H. (2023). Advances in xanthan gum-based systems for the delivery of therapeutic agents. *Pharmaceutics, 15*(2), 402.
89. Toth, I. K., Barny, M. A., Brurberg, M. B., Condemine, G., Czajkowski, R., Elphinstone, J. G., Helias, V., Johnson, S. B., Moleleki, L. N., Pirhonen, M., & Rossmann, S. (2021). Pectobacterium and Dickeya: Environment to disease development. In *Plant Diseases Caused by Dickeya and Pectobacterium Species* (pp. 39–84). Springer.
90. Pandey, P., & Prajapati, B. (2019). Soilborne microbes: A culprit of juvenile plants in nurseries. In *The Vegetable Pathosystem* (pp. 427–456). Apple Academic Press.
91. Fernandez-Piñeiro, I., Alvarez-Trabado, J., Márquez, J., Badiola, I., & Sanchez, A. (2018). Xanthan gum-functionalised span nanoparticles for gene targeting to endothelial cells. *Colloids and Surfaces B: Biointerfaces, 170*, 411–420.

92. Jacob, S., Nair, A. B., Shah, J., Sreeharsha, N., Gupta, S., & Shinu, P. (2021). Emerging role of hydrogels in drug delivery systems, tissue engineering and wound management *Pharmaceutics, 13*(3), 357.
93. Jog, R., & Burgess, D. J. (2018). Excipients used in oral nanocarrier-based formulations. In *Fundamentals of Nanoparticles* (pp. 279–342). Elsevier.

CHAPTER 6

Advances in Modified Drug Delivery Using Prosochit®

EMMANUEL O. OLORUNSOLA[1] and MICHAEL U. ADIKWU[2]

[1]*Department of Pharmaceutics and Pharmaceutical Technology, University of Uyo, Uyo, Nigeria*

[2]*Department of Pharmaceutics, University of Nigeria, Nsukka, Nigeria*

ABSTRACT

Prosochit® is a coprocessed excipient of prosopis gum and chitosan. While graft copolymerization is used to modify the properties of polymers with a minimum loss in their native properties, coprocessing of polymers goes further by harnessing the benefits of two or more polymers and minimizing the shortcomings of individual components. This chapter discusses Prosochit® in terms of its component polymers, process of development, and its exploration in modified drug delivery. Prosopis gum is obtainable from the seeds of plants in the genus *Prosopis.* It is a light-brown fluffy gum that possesses high bioadhesive strength, the value being comparable with that of sodium carboxymethylcellulose. It has been explored in both immediate and modified drug delivery. Chitosan, on the other hand, is normally obtained from the exoskeleton of aquatic animals such as crabs and prawns. Its purified form is available for pharmaceutical applications as diluents, binders, drug carriers, matrix former, viscosity modifiers, wetting agents, bioadhesive polymers, site-specific delivery excipients, dissolution enhancers, and permeation enhancers. Prosochit® in three forms had been derived by coprocessing of different proportions of prosopis gum and Callinectes chitosan. Prosochit®, now available as Prosochit® 201, 101, and 102, is being applied in the modified delivery of many drugs. It has been explored in the

Graft Copolymers of Polysaccharides for Drug Delivery Systems.
Rishabha Malviya and Sonali Sundram (Eds.)

delivery of artemether, a drug whose bioavailability is dissolution-limited. It has also been successfully utilized in the delivery of hydrochlorothiazide, a drug whose bioavailability is limited by low solubility and low permeability. Recently, the novel excipient was investigated in oral delivery of insulin. Preclinical studies have shown that the new excipient enables the oral delivery of this therapeutic agent, whose oral administration is highly hindered by gastric degradation and poor intestinal absorption. Therefore, Prosochit® is a promising excipient for oral delivery of medicaments having inherent delivery challenges.

6.1 INTRODUCTION

A polymer is a large molecule comprising many small repeating units, which are referred to as monomers. The monomers can be of the same kind or of different types. They may be olefinic or functional with reactive groups. The types of monomers involved determine the polymerization method to be utilized in the formation of the polymer [1]. The name given to a polymer depends on the number of monomers involved. It could be dimer, trimer, tetramer, pentamer, and so forth [1]. An oligomer contains about 100 monomers, while the term "polymer" is often reserved for products containing more than 200 monomeric units. Polymers are generally characterized by high molecular weights, and their chemical compositions vary greatly [2]. They display different behaviors depending on their structure, linearity, molecular mass, and molecular interactions. Hence, they provide various mechanisms for the delivery of many complex and new medicinal agents. Biodegradable and bioabsorbable polymers are extensively utilized for drug delivery purposes [3, 4].

Based on their origin, polymers can be classified into three, namely: natural, semi-synthetic, and synthetic polymers [5]. While natural polymers are obtained from plants and animals, synthetic polymers are formed by chemical reactions. Synthetic polymers like the acrylates or methacrylates can be classified into cationic, anionic, and neutral (nonionic) polymers. Eudragit E100 is a cationic acrylate containing less than 3000 ppm of residual monomers. Anionic polymers such as Eudragit L100 and S100 contain methacrylic groups, which only dissociate at the various higher pH of the intestine, enabling drug delivery to the regions with the corresponding pH. The anionic polymers are employed as film-formers in enteric coating. Neutral acrylate or methacrylate with or without aminoethyl functionality are generally insoluble or swellable. They are used for sustained release where insolubility is very much desirable. Eudragit RS100 is an example of neutral synthetic polymers. Semisynthetic polymers are those that are derived from nature but have been subjected to chemical processes to improve their quality. An example of semisynthetic polymers is cellulose acetate [4].

While graft copolymerization is used to modify the properties of polymers with a minimum loss in their native properties, coprocessing of polymers is used to harness the benefits of two or more polymers and to minimize the shortcomings of individual components. Prosochit® is a coprocessed excipient of prosopis gum and chitosan [6]. The component polymers are of natural origin and have been severely investigated individually for drug delivery purposes.

6.2 PHARMACEUTICAL POLYMERS

These are polymers that are employed for drug delivery purposes. They serve various functions in the preparation and performance of pharmaceutical formulations [7]. They can be utilized in the preparation of both immediate-release as well as modified-release formulations.

Pharmaceutical polymers can be classified into water-soluble and water-insoluble polymers [2]. Cellulose ethers with methyl and hydroxypropyl substitutions are soluble in water, while ethyl cellulose and a group of cellulose esters are soluble in organic solvents. Natural gums are either water-soluble or swell up, forming mucilage. They are insoluble in alcohol and ether. Synthetic polymers such as polyethylene glycol, polyvinyl pyrrolidone, and polymethacrylate esters containing anionic and cationic functionalities are soluble in water and are widely employed for pharmaceutical use. In another classification, which is based on the chemical nature, pharmaceutical polymers can be classified into starch and starch derivatives, cellulose and cellulose derivatives, gums, and chitosan [8]. Polymers in these different classes are employed for different pharmaceutical uses, and having the knowledge of the chemical class to which a polymer belongs can serve as a pointer to its applicability.

6.2.1 STARCH-BASED POLYMERS

Starch can be derived from roots and tubers, with common sources including corn, rice, millet, yam, potato, and cassava. It is widely used as a diluent, binder, and disintegrant in tablet formulations, either in its native or modified form. Co-processing with other polymers has been employed to enhance starch's properties. For example, pregelatinized starch–polyvinyl pyrrolidone is produced by gelatinizing potato starch in the presence of polyvinyl pyrrolidone [9]. Starch properties can also be improved by heating, drying, and treating it with hydrochloric acid to produce dextrin.

6.2.2 CELLULOSE-BASED POLYMERS

Cellulose is known to be the most abundant natural polymer. Even though it is not soluble in water, its water-soluble derivatives have been widely applied in the formulation of pharmaceutical products [8]. Methylcellulose, carboxymethylcellulose, and hydroxypropyl methylcellulose (HPMC) are the cellulose derivatives most commonly used for pharmaceutical formulations. When the temperature of a solution of any cellulose-based polymer is increased, the hydrophobic group of the polymer starts to aggregate. This leads to the formation of a cloudy system. The cloud point temperature for pure methylcellulose, which has no hydroxypropyl substitution, for instance, is about 50°C. Substitution of the methyl group with the hydroxypropyl group, which is more soluble in water, results in an increase in the cloud point temperature. The number of the methyl group substituted determines the extent of the increase in the cloud point temperature. In general, the value of the cloud point temperature depends on the extent of substitution of the methyl groups, while the viscosity of the solution depends on the molecular weight of the polymer.

6.2.3 GUMS

Gums are colloidal substances with high molecular weights [10]. They can also be defined as water-soluble or water-swellable hydrocolloids which are insoluble in organic solvents. Gums are generally characterized by low toxicity and high biodegradability, and are compatible with many drugs. Natural gums are formed by natural processes. They are obtained from living organisms, and their sources include plant exudates, plant extracts, plant seeds, seaweed extracts, micro-organisms, and animals. They are highly susceptible to microbial attack. This is the main drawback of their application as pharmaceutical excipients [11].

a. **Plant Exudate Gums:** Exudate gums are obtained as patches of gluey substances on the surface of some plant barks [11]. They occur as a result of injuries inflicted on susceptible plants. The injury may be caused by an adverse weather condition, microbial attack, insect attack, or any form of mechanical damage. Gum–producing trees do not exude gum when accidentally scorched by fire, probably due to the destruction of *Bacterium acacia* and enzymes. However, if re-infection occurs, gum is produced, implying that bacteria play a major role in gum formation. Some examples of plants that produce gum as exudates are acacia, tragacanth, Gatti, cashew, and khaya.

b. **Plant Extract Gums:** These are polysaccharides that are extracted from citrus fruits [12]. Such polysaccharides are called pectins and are derived from ripening product of the peels of green fruits. They are formed by the conversion of protopectin to pectinic acid and then to pectin. Pectins are predominantly linear polymers of (1–4)-linked D-galacturonic acid residues interrupted by 1,2- linked L-rhamnose residues with a few hundred to one thousand building blocks per molecule [12]. Hence, the major component is galacturonic acid.

c. **Seed Gums:** These gums are extracted from seeds. Examples are prosopis, afzelia and psyllium. They are more complex in structure compared to the exudates gum. They also have lower solubilities compared to exudates gums. They are better solubilized in dilute alkaline solutions; and are often referred to as hemicelluloses [13].

d. **Seaweed Gums:** These are gums that are obtained as extracts from seaweed. They include alginates, agar, and carrageenans. Alginates are natural polysaccharides obtainable from the seaweed (*Phaeophyceae*). They exist as linear polymers comprising the D-mannuronic acid and L-glucuronic acid residues. Alginates are the salt forms of alginic acid, and sodium salt is the most common form for application in drug delivery. They are used in the formulation of the matrix-type beads, in the formulation of liposomes, and in the modulation of GI residency.

   Gum agar, as an extract, is obtained from the marine algae of the class *Rhodophyceae*. Such algae are commonly found in the waters along the coast of Japan, New Zealand, Mexico, South Africa, Portugal, and Morocco [12]. Carrageenans are sulfated polysaccharides of galactose and anhydro-galactose, which are arranged linearly. They can be obtained as extracts from the red algae in the *Florideophyceae* class [14–16]. Carrageenan is characterized by a half-ester sulfate group, which has the ability to react with proteins in solution, forming a gel or the solution becoming viscous. The incorporation of a carrageenan to a solution containing proteins causes gelling or increase in viscosity as a result of complexation of the carrageenan with the charged amino acids of the protein. Hence, a formulator must be vigilant of such incompatibility to ensure the stability of the drug formulation.

e. **Microbial Gums:** Microbial gums are products of the fermentation process. Examples of such are Xanthan, Gellan, and Pullulan. Xanthan gum (XG) is a product of glucose fermentation carried out by *Xanthomonas campestris.* It is a microbial exopolysaccharide made up of *D-glucosyl, D-mannosyl*, and *D-glucuronyl* acid residues as the major components, which are also present in a molar ratio of 2:2:1. Pyruvyl and acetyl residues may also be present in variable proportions. This gum is acidic in nature and contains pentasaccharide subunits. Gellan is another type of microbial gum. It is secreted by *Sphingomonas elodea.* It is anionic in terms of charge and consists of repeating tetrasaccharide units of glucose, glucuronic acid, and rhamnose residues in the proportion of 2:1:1. Pullulan is a microbial gum produced by many species of the fungus *Aureobasidium, specifically the A. pullulans [12].* It is a homo-polysaccharide of maltotriose units. Each maltotriose unit contains three glucose units.

f. **Animal Gums:** These are referred to as gelatin. They are normally derived from animal skin and bones. Gelatin is cationic, carrying a net positive charge, in contrast to plant gums, which are mostly nonionic [17]. Like other natural gums, batch-to-batch inconsistency and susceptibility to microbial attack are the major limitations of gelatin as a pharmaceutical polymer.

6.2.4 CHITOSAN

Chitosan is a natural polymer of animal origin. It is derived from chitin and found in the exoskeleton of aquatic animals such as crabs and lobster [18]. It is cationic in terms of charge and possesses mucoadhesive properties. Chitosan has been investigated as an excipient for controlled drug delivery and for the formulation of mucoadhesive dosage forms because of the gelling and adhesive properties, respectively. It can be used as a diluent, disintegrant, binder, drug carrier or as a platform for parenteral formulations [19]. It is also useful as an absorption enhancer to promote the paracellular permeability of drugs [20]. Chitosan can be mixed with non-ionic surfactants such as sorbitan esters for use in the formulation of emulsion-like solutions and creams. Chitosan matrix treated with an anionic compound such as pentasodium tripolyphosphate (TPP) is used for preparing chitosan beads or microspheres [21].

6.3 APPLICATIONS OF PHARMACEUTICAL POLYMERS

The properties of polymers that make them to be useful as pharmaceutical excipients are film formation, thickening, gelling, adhesion, barrier property, pH-dependent solubility, and solubility in organic solvents. These characteristics are necessary for coating, viscosity enhancement, controlled release, binding, protection, targeted delivery, and taste masking. Polymers are widely applied in solid, semi-solid, and liquid formulations [22]. The natural ones especially have found wide usage in Pharmacy because of their low cost, non-toxicity, biodegradability and several other appropriate physicochemical characteristics [10].

6.3.1 AS EXCIPIENTS IN SOLID DOSAGE FORMS

In addition to their roles in the manufacturing process, pharmaceutical excipients may be involved in transporting the medicament to the site of action in the body. They may also be incorporated to enable drug release at a definite part of the gastrointestinal (GI) tract. Others are incorporated to enhance product performance, such as promoting the disintegration or dissolution of the formulation or dissolution of the drug [7]. Some others are added to improve patient compliance (especially in children) simply by improving taste and appearance. Polymers as excipients in solid dosage are useful as diluents, binders, disintegrants, controlled delivery agents, film formers, and so on.

a. **Diluents:** Diluent or fillers are inert ingredients that are used to create the desired bulk, improve the flow properties of capsule formulation, or improve the flow properties and compression characteristics of tablet formulation. The diluent selected should be such that the product stability is not adversely affected. Examples of diluents include: lactose, mannitol, microcrystalline cellulose, sorbitol and starch and chitosan [23].

b. **Binders:** These are adhesives added to tablet formulations to bring cohesiveness to the powders, thereby providing the necessary bonding for granule and compact formation. Examples include acacia gum, alginic acid, gelatin, povidone, pregelatinized starch, carboxymethylcellulose, and microcrystalline cellulose. Binders can be added to powders in three different ways. They may be added as a dry powder (e.g., pregelatinized starch and polyvinyl pyrrolidone) mixed with other ingredients before wet agglomeration [24]; as a solution or mucilage of the binder (e.g., all native starch) which is used as agglomeration liquid [25] or as a powder (e.g., microcrystalline cellulose, microcrystalline starch) mixed with other ingredients followed by direct compression [23].

c. **Disintegrants:** Disintegrants, after drug administration, facilitate the breakup of tablets into smaller particles that are more readily dispersed or dissolved [26]. They can be added just before granulation or, during granulation, or during the lubrication step prior to compression. They may also be added at once or in divided portions incorporated at different processing steps. Examples of disintegrants include alginic acid, microcrystalline cellulose, and methylcellulose. Disintegrants, based on their chemical composition, have been classified into starches, dyes, cellulose, algins, gums, and cross-linked polymers. Corn and potato starch are the initial and most common disintegrants. Starch in the proportion of 5% of tablet weight has been suggested for use as a disintegrant. If more rapid disintegration is desired, the amount may be increased to 10 or 15%. Certain agents, called super-disintegrants, are effective at concentrations below 5%. Examples are croscarmellose, crospovidone, and sodium starch glycolate [27].

It has been postulated that disintegrants work by three different mechanisms. These are swelling, capillary action, and carbon dioxide production [27]. Esezobo et al. [28] proposed that disintegrants in the presence of water swell and burst open. This is the swelling theory, and starch is thought to work with this method. For capillary action, the tablet, in an aqueous environment, takes up water, leading to a build-up of internal pressure, which eventually results in the breakage of the tablet. Cross-linked polyvinylpyrrolidone works by this method. The third mechanism is the production of carbon dioxide when the formulation comes in contact with water [29, 30].

d. **Controlled Delivery Agents:** Polymers are extensively utilized to achieve controlled drug delivery, to enhance stability, and to increase bioavailability [31–33]. A formulation can be such that the drug is dispersed within a polymer matrix or that the polymer is used as a membrane or coating. In a matrix system, the rate of drug release depends on the initial concentration of the drug and the extent of relaxation of the polymer chains. The extended-release alprazolam tablet is an example of a matrix system in which prolonged drug release is achieved by swelling and erosion of the matrix [34].

   For the membrane systems, the polymer system may be porous or non-porous. Drug release through a non-porous membrane is basically driven by diffusion, while a porous membrane generates an extra route for the drug release, which is through the pores. The medicament from the drug core is normally transferred across the membrane. The extent and rate of drug release from membrane systems can be controlled by altering the membrane thickness, varying the amount of plasticizer incorporated, and modifying the structure of the pores [35]. Membrane systems have been applied to enhance drug stability and formulate enteric release and sustained release tablets. Enteric-coated products are able to pass through the stomach safely so that the drug can be released in the intestine where the pH is higher. In such cases, the preparation involves coating the formulation with a polymer which is pH-sensitive [36]. An example of such polymer is Eudragit L100 which is anionic in nature.

e. **Film Formers:** Ability of polymers to form films is an important factor in the formulation of breath films, cough strips and sore throat strips. In general, linear gums have more sites available for intermolecular hydrogen bonding. Hence, they are better film formers. The seaweed gums such as agar, alginate and carrageenan as well as their blended products are useful in the formulation of film dosage forms [4].

   Polymers are also used for coating of tablets. They can be used individually or as blends. A blend of hydroxypropyl cellulose

(HPC) with HPMC has been prepared to improve the film adhesion of the latter. HPC has also been found to reduce film defacing, deformation and cracking when combined with HPMC [37]. Tablet coating may be for functional or non-functional purposes. The non-functional coatings are normally employed for the purpose of esthetics, improvement of stability to humidity and light, product identification, and/or taste masking. HPMC is the most commonly used polymer for non-functional coating. When this polymer is used for tablet coating, the addition of a plasticizer is required to enhance the film flexibility. Polyethylene glycol as plasticizer is usually combined with HPMC for use as a coating material [4].

Polymer blends have been applied extensively as functional film coatings for controlled release purposes [38]. Aqueous systems are preferred to organic solvent-based solutions for the application of coatings [39]. Coatings, usually, are colloidal dispersions of water-insoluble polymers. Examples of such polymers are: ethylcellulose, polyvinyl acetate and the polymethacrylates.

6.3.2 AS EXCIPIENTS IN SEMI-SOLID DOSAGE FORMS

In semi-solid formulations, polymers are used as emulsifying agents in creams. They are also used as gelling agents [40]. Chitosan and ovalbumin protein have been recommended for use as gel formers in the manufacture of pharmaceutical and cosmetic semi-solids.

6.3.3 AS EXCIPIENTS IN LIQUID DOSAGE FORMS

Apart from their extensive use in the manufacture of solid formulations, polymers have been widely applied as viscosity modifiers in the manufacture of liquid dosage forms. They are also incorporated into liquid systems for the control of viscosity of aqueous solutions and for stabilization of suspensions and emulsions. Most polymers exhibit thickening activity, while some others manifest gelling activity. Thickening is desired for suspensions and emulsions while gelling is applicable to semi-solid formulation and also for utilization in controlled drug delivery. Guar gum (GG) and the likes are characterized by excellent thickening property while alginates and chitosan exhibit good gelling property [11].

a. **Solubilizing Agent:** Different gums in aqueous solutions behave in different manners. For instance, concentrated XG solution resists flow due to excessive hydrogen bonding, but manifests shear-thinning behavior on application of shear stress. This behavior is critical in the manufacture of pharmaceuticals and cosmetics. The work of Olorunsola and John [41] has shown that cashew gum (CG) has a better solubilizing property compared to acacia gum.

b. **Emulsifying Agent:** Emulsifying property is a unique feature of Gum Arabic (acacia gum). Locust bean gum also possesses a good emulsifying property. Like many other gum solutions, locust bean gum displays shear-thinning flow. Afzelia gum from *Afzelia africana* seed has been found to possess appreciable emulsifying properties when investigated in liquid paraffin emulsion formulation [42].

c. **Suspending Agent:** XG has been employed as a suspending agent in some liquid formulations including cefdinir oral suspension. It is very effective in the stabilization of liquid dosage forms at acidic pH. Its application in liquid formulation is based on the thickening property. Due to the rigid structure, aqueous solution of XG is very stable over a wide pH range. *Raphia africana* hydrocolloid has also shown a positive result when investigated for suspension formulation [43].

6.4 MODIFICATION OF POLYMERS FOR ENHANCED PHARMACEUTICAL USE

In situations where a polymer cannot meet a particular need, its properties can be modified to obtain a better substance. For instance, drug release in the colon can be enhanced if the hydrophobicity of the polymer used for the delivery is modified. The characteristics of a polymer can be changed simply by modifying its chemical nature. For example, while natural chitosan is only soluble at low pH, its carboxymethyl derivative is soluble at a wider pH range [21]. Multiple polymers can also be blended together or chemically reacted. With blending, a multiphase system generally results, whereas with chemical reaction, a monophase system is formed. The distinction can clearly be investigated using differential scanning calorimetry (DSC) by monitoring the glass transition temperature (*Tg*) of the individual polymers. For blended polymers, more than one *Tg* value will be seen, while only one will be seen in situations of chemical reactions [44].

6.4.1 COPOLYMERIZATION

Copolymerization is a form of polymerization whereby more than one type of monomer is involved. If only one type of monomer is involved in the reaction, the process is called polymerization, and the product becomes a homopolymer. For example, polyethylene is a homopolymer of ethylene. Copolymerization, on the other hand, involves different types of monomers, and the product is referred to as a copolymer. Graft copolymerization is a form of polymerization whereby one type of monomer, known as the side chain, is incorporated into another polymer chain comprising another monomer type. Hence, a copolymer has a polymer chain as a backbone and another chain of polymer as a side-graft. Copolymers have many benefits both from the graft and for the backbone, and graft copolymerization is used to modify the properties of polymers with minimum loss in their native properties [45].

6.4.2 CO-PROCESSING OF POLYMERS

Coprocessing is the sub-particulate interaction of two or more substances leading to overall improvement in the functionality of the product [46]. Hence, polymer coprocessing involves the processing of two or more polymers in a molten or solution state. The resulting substance is thereafter solidified at lower temperature if the preparation involved melting or recovered at higher temperature if the preparation involved dissolution. Coprocessing should not be confused with polymer blending which means simple mixing of two or more polymers [47]. Coprocessing of polymers produces granulates having superior characters compared with the simple mixture of the components or with the individual component. Co-processed excipients can be developed to address issues such as flowability, compactability, or disintegrant effect. Combination of the polymers should be such that the selected polymers complement each other to mask the undesirable characters of the individual excipient and also retain or improve the desirable characters [48].

Co-processing is another means whereby new excipients are developed without going through the rigorous safety testing of a completely new substance [49]. So, it can be defined as the combination of two or more established excipients by an appropriate method. Examples of co-processing methods are co-drying, solvent extraction, co-precipitation, spray drying, melt granulation, gelatinization, freeze-thawing, and kneading [9]. The development of Prosochit® involves co-precipitation of Prosopis gum and Callinectes chitosan followed by drying [6].

6.5 PROSOPIS GUM

6.5.1 BIOLOGICAL SOURCE

Prosopis gum is derived from the seeds of *Prosopis Africana*. The plant has an open crown and a slightly rounded buttress. It also has dropping foliate alternate bipinnate leaves reaching a length of 10–15 cm. The flowers are cream to yellow in color, fragrant, and dense with 6–10 cm long axillary spikes. The fruit is used for feeding animals, while the seeds are used for making locust beans [50].

6.5.2 GEOGRAPHICAL SOURCE

Prosopis Africana is commonly found in the tropics. It is the only Prosopis specie that is found from the extreme of West Africa to the extreme of East Africa. It is therefore, native to virtually all the countries in West and East Africa. In Nigeria, it is in abundance between latitude 7° N and 10° N [50].

6.5.3 CHEMICAL PROPERTIES

Prosopis gum is a natural highly branched polysaccharide having a chain-like structure. It consists mainly of glucose, galactose, fructose, and xylose as the monosaccharide units; and can be described as a xyloglucan or xyloglucan [13].

6.5.4 PHYSICAL PROPERTIES

The gum is light brown in color and fluffy in texture. It is poorly soluble in water but solubilized in aqueous alkali. Its ability to lower the surface tension of water is comparable to that of sodium carboxymethylcellulose [51].

6.5.5 PHARMACEUTICAL USES

Prosopis gum has been largely investigated for the formulation of both immediate-release and modified-release systems. In the formulation of the immediate-release tablet of artemether, it was observed that tablets containing Prosopis gum had higher crushing strength, higher friability, and longer disintegration time compared to those containing CG at corresponding binder concentrations. It was also observed that tablets containing 3%w/w CG as the binder brought forth complete drug release within 1 h while those containing the same concentration of prosopis gum brought forth less than 75% drug release over the same period of time. It was concluded that CG is suitable for use as a binder over a relatively wide range of concentrations to achieve immediate drug release while prosopis gum enables prolonged or delayed drug release [52].

In another work involving the use of Prosopis gum as a tableting excipient, it was observed that tablets containing 4%w/w Prosopis gum enabled cumulative drug release of 27.2%, 57.0%, and 78.6% in 30 min, 60 min, and 120 min, respectively. It was concluded that Prosopis gum at that concentration could cause a delay in the delivery of artesunate such that 57 to 79% of the drug is released in the stomach while 21 to 43% is released in the intestine [53].

6.6 CALLINECTES CHITOSAN

6.6.1 BIOLOGICAL SOURCE

The exoskeleton of members of the genus *Callinectes* is a good source of chitosan. *Callinectes* is a genus of crabs that contains 16 species. The work of Olorunsola et al. [27] has demonstrated the presence of valuable chitosan in the exoskeleton of *Callinectes gladiator*.

6.6.2 GEOGRAPHICAL SOURCE

Crab is part of the diet of the coastal towns of West African countries including Nigeria, Benin, Togo, Ivory Coast, and Ghana [27, 54]. *Callinectes gladiator* is a decapod crustacean which is abundant in the coastal areas of these countries [27].

6.6.3 PHYSICAL PROPERTIES

Chitosan is a derivative of chitin and it's sometimes referred to as soluble chitin. While chitin is practically insoluble in water, chitosan is water-soluble. The work of Olorunsola et al. [19] has shown that chitosan has better flow properties compared to sodium carboxymethylcellulose.

6.6.4 CHEMICAL PROPERTIES

Chitosan's structure consists of randomly arranged β-(1–4)-linked D-glucosamine deacetylated units and N-acetyl-D-glucosamine units [55]. The amino group in chitosan has an acid dissociation constant (pKa) of 6.5. The polymer is both biocompatible and biodegradable [19].

6.6.5 PHARMACEUTICAL USES

A previous work reported that chitosan incorporated as an absorption promoter at concentrations of 1–5% influenced the disintegration of ciprofloxacin tablets in a positive manner [20]. Other possible pharmaceutical uses of chitosan include [19]:

- Diluent in direct compression tablet
- Binder in wet granulation
- Modified-release matrix formation
- Drug carrier in microparticulate system
- Film former in controlled drug release
- Preparation of hydrogels
- Viscosity modifier
- Improvement of dissolution of poorly soluble drugs
- As a bioadhesive
- For targeted drug delivery
- Biodegradable polymer in implants
- Carrier in vaccines, peptides, and protein delivery.

6.7 DEVELOPMENT OF PROSOCHIT®

The development of a co-processed excipient must start with the identification of suitable substances to be combined, the appropriate proportions to be used, and the preparation method to be employed to get an optimized product with desired physicochemical properties. It could involve the selection of an excipient of reasonable price in combination with optimal proportion of a functional material such that a product with superior functionality compared to the simple mixture of the components will be produced [47]. Prosochit® is a novel co-processed excipient derived from prosopis gum and Callinectes chitosan [6]. It is of three types coded as Prosochit® 201 (PC201), Prosochit® 101 (PC101) and Prosochit® 102 (PC102). The three types were developed by coprocessing prosopis gum and Callinectes chitosan in the proportions of 2:1, 1:1, and 1:2, respectively. The procedure involves the three steps of dispersion formation, precipitation, and drying [6].

6.8 COMPATIBILITY OF PROSOPIS GUM WITH CALLINECTES CHITOSAN

A wide variety of properties are achievable when chitosan is blended with other multifunctional polymers [56]. This is the basis for coprocessing of prosopis gum and Callinectes chitosan. There shouldn't be any adverse interaction between the polymers being used so that the product will be of acceptable quality. A study has been carried out using FTIR and DSC to determine the compatibility of prosopis gum with Callinectes chitosan [57]. The obtained FTIR spectra of the individual polymers (Prosopis gum and Callinectes chitosan) and those of the three co-processed excipients (Prosochit® 201, 101, and 102) are shown in Figure 6.1.

For Prosopis gum, the absorption peaks observed at positions 3443.00, 2928.14, 1639.78, and 1452.84, 1390.15, 1085.23, and 1028.75 cm^{-1} were linked to aldehydes and alcohols, while those at 871.50 and 803.10 cm^{-1} were ascribed to the presence of cyclic molecule [57]. For chitosan, the broad bands at 3240.00 and 1495 cm^{-1} were ascribed to N-H stretching and bending of amino and acetamido groups, respectively, and those at 1062.43 and 1013.40 cm^{-1} were ascribed to O-H of alcohol while those at 862.54, 786.01 and 700.51 cm^{-1} were ascribed to C-H bending of a cyclic molecule. In addition, the peaks between 2300 and 2600 cm^{-1} observed in the spectra of Prosopis gum and Callinectes were linked to the presence of carbon dioxide impurities [57]. Examination of the spectra of the three types of Prosochit® has shown that all the major absorption peaks and the bands observed in the spectra of Prosopis gum and Callinectes chitosan were seen in the three spectra of Prosochit® showing that the two-component polymers are compatible. The unordered peaks and those that were ascribed to impurities in the component polymers were not found in the spectra of Prosochit®, which indicates the effectiveness of the coprocessing [57].

Differential scanning thermograms of Prosopis gum, Callinectes chitosan, and the three types of Prosochit® were also studied by Olorunsola and Usungurua [57]. The thermogram of Prosopis gum was observed to be characterized by an exothermic peak at 310°C, while that of Callinectes chitosan was characterized by two endothermic peaks at 129 and 179°C. On the other hand, Prosochit® 201, Prosochit® 101, and Prosochit® 102 were characterized as having an endothermic transition each at positions 180°C,

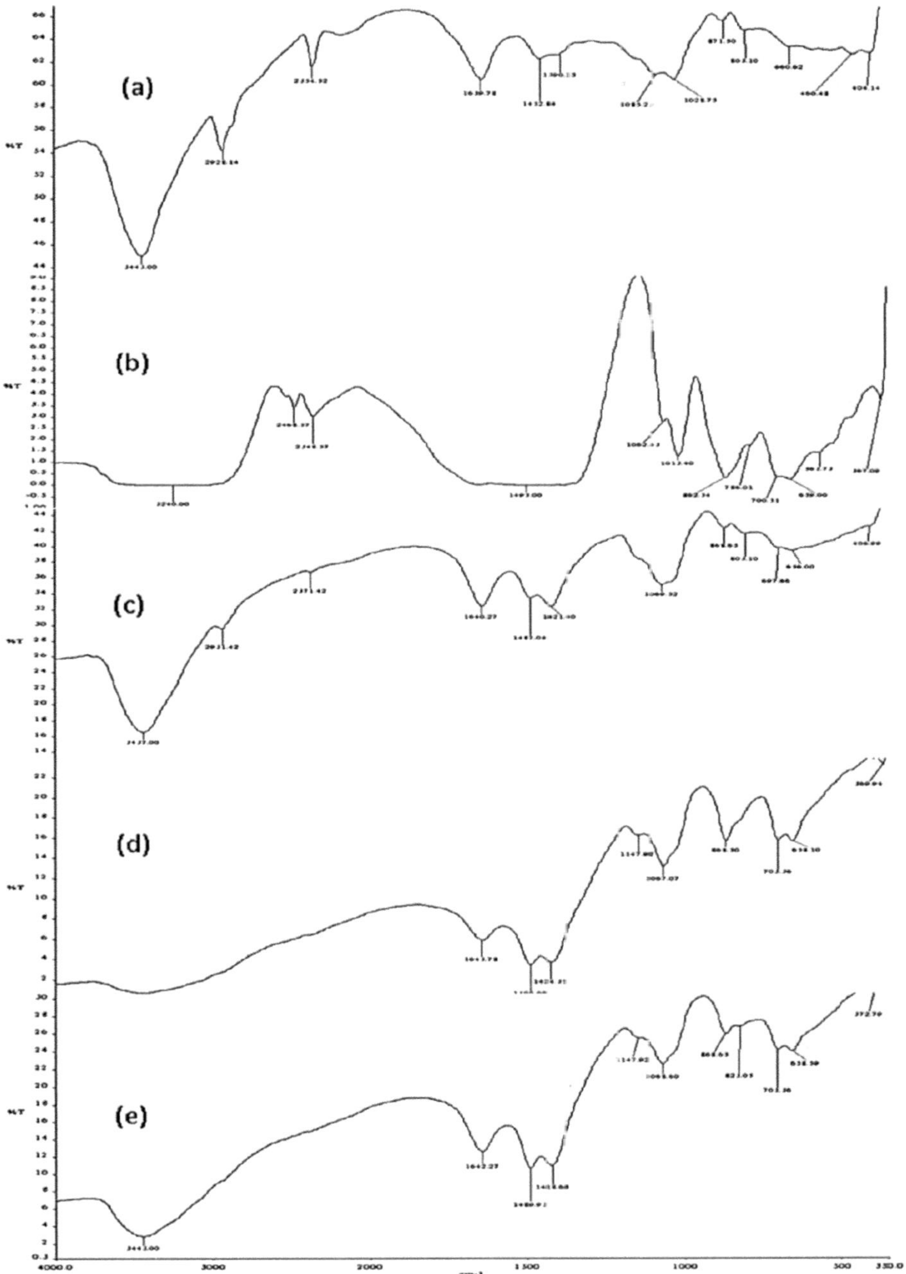

FIGURE 6.1 Fourier transform infrared spectra of (a) Proscpis gum; (b) Callinectes chitosan; (c) Prosochit® 201; (d) Prosochit® 101; and (e) Prosochit® 102.

Source: Adapted from Ref. [57]

179°C and 178°C, respectively. While the exothermic transition of Prosopis gum was linked to polymer degradation based on the work of Iqbal et al. [58], the first and second endothermic transitions in the thermograms of chitosan were linked to enthalpy relaxation and melting, respectively, based on the description given by Hovat et al. [59]. It was concluded that the melting point of Prosochit® is close to that of pure chitosan and that the value slightly decreases with an increase in the proportion of chitosan [57].

6.9 CHARACTERIZATION OF PROSOCHIT®

6.9.1 DENSITIES

The various density measures of Prosopis gum, chitosan, and the three types of Prosochit® as determined in the previous work of Olorunsola and Usungurua [57], are shown in Table 6.1. The bulk densities of prosopis gum and Callinectes chitosan are not significantly different, while those of the three types of Prosochit® are significantly lower. It was found that the co-processing led to a decrease in the bulk density of the excipients. The same conclusion was made of the tapped density. For the true density, the values were adjudged to be significantly different comparing Prosopis gum with Callinectes chitosan, while the value for Prosochit® was found to decrease with an increase in the proportion of chitosan used.

TABLE 6.1 Polymer Densities and Flow Properties [57]

Polymer	Bulk Density (g/mL)	Tapped Density (g/mL)	True Density (g/mL)	Hausner's Ratio	Carr's Index (%)
Prosopis gum	0.71±0.00	0.84±0.01	1.55±0.00	1.19±0.01	15.77 ± 1.31
Chitosan	0.81±0.01	0.94±0.01	1.06±0.00	1.15±0.02	13.50 ± 1.16
Prosochit® 201	0.55±0.00	0.66±0.01	1.31±0.00	1.19±0.02	16.22 ± 0.84
Prosochit® 101	0.53±0.00	0.64±0.01	1.24±0.00	1.19±0.02	16.16 ± 1.59
Prosochit® 102	0.53±0.01	0.63±0.01	1.14±0.00	1.17±0.01	14.78 ± 1.17

Data are mean ± SEM.

6.9.2 FLOWABILITY

The flow parameters of the excipients, as determined by Olorunsola and Usungurua [57], are also shown in Table 6.1. The Hausner's ratio of the various substances is not significantly different. Similarly, the values of the Carr's index of the excipients are not significantly different. Hausner's ratio and Carr's index are used as indirect measures of powder flowability. Hausner's ratio of less than 1.25 or Carr's index of less than 20% indicates good flow; Hausner's ratio of 1.25 to 1.50 or Carr's index 20 to 33% indicates fair flow, while Hausner's ratio above 1.5 or Carr's index greater than 33% indicates poor flow [60]. Thus, all five excipients are characterized by good flow.

6.9.3 SWELLABILITY

The swellability of the three types of Prosochit® has also been studied [57]. The swelling index ranges from 500 to 720%. It is greater than that of chitosan and lower than that of Prosopis gum. The value decreases with an increase in the proportion of Callinectes chitosan included in the preparation. Swellability is an important factor in controlled drug delivery [61]. Like Prosopis gum, Prosochit® is highly swellable, and the various types of Prosochit® may be investigated for controlled delivery of drugs requiring such technology. Prosochit® has an added advantage of permeation-enhancing property due to the presence of chitosan. This is in agreement with the work of Ngwuluka et al. [46], which stated that a combination of two polymers for application as a swellable matrix is superior to the use of the individual ingredient.

6.9.4 CHARACTERISTICS OF THE DISPERSIONS

The physical characteristics of 2%w/v dispersion of the excipients, as reported in previous studies [57], are presented in Table 6.2. The pH of Prosochit® increased with an increase in the proportion of chitosan used in the coprocessing since chitosan has a higher pH compared to Prosopis gum. There is a significant difference in the viscosity of 2%w/v dispersion of the various excipients, with the value decreasing with an increase in the proportion of chitosan. Hence, an increase in the proportion of chitosan will likely reduce the effectiveness of Prosochit® as a suspending or emulsifying agent. There is no significant difference in the adhesive strength of the five excipients. However, Prosochit® had a higher value compared to the individual component. The observation is attributable to synergism of the bioadhesive effects of the two components [62, 63]. Hence, the co-processing of the two polymers leads to an increase in the adhesive strength.

TABLE 6.2 Characteristics of 2%w/v Dispersions of Some Excipients [57]

Polymer	pH	Viscosity (mPas)	Adhesive Strength (Nm^{-2})
Prosopis gum	8.31 ± 0.01	7.18±0.00	13.80±1.48
Chitosan	10.66 ± 0.04	2.88±0.04	13.26±0.24
Prosochit® 201	10.13 ± 0.02	4.47±0.04	15.12±0.80
Prosochit® 101	10.16 ± 0.00	3.89±0.04	14.59±0.96
Prosochit® 102	10.22 ± 0.03	3.26±0.02	14.06±0.27

Data are mean ± SEM.

6.10 EXPLORATION OF PROSOCHIT® FOR ORAL DELIVERY OF ARTEMETHER

Artemether is a low solubility – high permeability drug [64]. Hence, its bioavailability is limited by the dissolution rate. A previous study has shown the compatibility of Prosopis gum with artemether [65]. Another study was carried out to determine the suitability of Prosochit® for enhanced oral delivery of this drug [64]. Each of the three types of Prosochit® was used as a wet granulation binder at concentrations equivalent to 2% w/w Prosopis gum for the formulation of artemether granules. The granules were subsequently tableted, and the compacts were assessed for mechanical strength, disintegration time, and dissolution rate. It was observed that there was no significant difference in the crushing strength of the tablets, but the friability of tablets containing mere Prosopis gum was significantly higher than that of tablets containing Prosochit®. The disintegration times were 67, 63, 52, and 38 min for tablets containing Prosopis gum, Prosochit® 201, Prosochit® 101, and Prosochit® 102, respectively, and the percent drug release at 1 h was in the inverse order Prosopis gum < Prosochit® 201 < Prosochit® 101 < Prosochit® 102. The tablets containing mere Prosopis gum and Prosochit® 201 failed the dissolution test, while those containing Prosochit® 101 and Prosochit® 102 passed the test.

The drug release from tablets containing Prosochit® 101 and Prosochit® 102 being better than those containing the other two excipients was linked to the presence of the adequate amount of chitosan in those excipients as the polymer is known to enhance drug release as already explained by Sonia and Sharma [21], Olorunsola et al. [27] and Sinha et al. [66]. Chitosan also possesses disintegrant property [27]. The ability of Prosochit® to enhance the oral delivery of artemether is also in agreement with the report of Ngwuluka et al. [46], which reported the superiority of co-processed excipients or excipient blends over the individual components. It was concluded that the flowability of the artemether granules and the mechanical properties of the tablets formulated with Prosochit® are comparable with those of Prosopis gum. It was also concluded that Prosochit® is better than Prosopis gum for the delivery of artemether and that Prosochit® 102 offers the best delivery of the drug based on the various formulations studied. The performance of the formulation containing Prosochit® 102 is ascribable to the relatively high amount of chitosan in the artemether tablet formulation.

6.11 EXPLORATION OF PROSOCHIT® FOR ORAL DELIVERY OF HYDROCHLOROTHIAZIDE

Hydrochlorothiazide is a low solubility – low permeability drug; thus, belonging to the Biopharmaceutics Classification System (BCS) class IV [67]. Different types of Prosochit® are characterized by different adhesive and release properties. In a study to investigate the influence of the various types of Prosochit® on the release and permeation of hydrochlorothiazide, the compatibility of the new excipient with the drug was studied using FTIR Spectroscopy, and six different tablet formulations were prepared using the wet granulation method; formulations F1 to F3 containing 3% PC201, PC101 and PC102, respectively and formulations F4 to F6 containing 6% PC201, PC101 and PC102, respectively. Tabet dissolution and drug permeation through the intestine were also studied in comparison with a marketed product [68].

Comparison of the Fourier transform infrared spectra of hydrochlorothiazide, Prosochit®, and the drug-excipient mixture revealed no adverse interaction between the drug and Prosochit®. The cumulative percent drug release in 1 h was also found to be in the order marketed formulation > F1 > F2 > F6 > F3 > F5 > F4 (Figure 6.2). The test formulations were also found to be characterized by better drug permeation compared to the marketed product, with Batch F1 being characterized by a markedly high permeability (Figure 6.3).

The work has shown that in spite of Prosopis gum and Callinectes chitosan (the two components of Prosochit®) being known to be good binders, they are capable of enhancing drug absorption. Prosopis gum possesses mucoadhesive properties, while chitosan promotes drug permeation through the paracellular route [20, 62]. It was concluded in that work that of the Prosochit® type/concentration investigated, 3% Prosochit® 201 has the most pronounced effect on the improvement of the dissolution and permeation of hydrochlorothiazide, a drug in the BCS class IV.

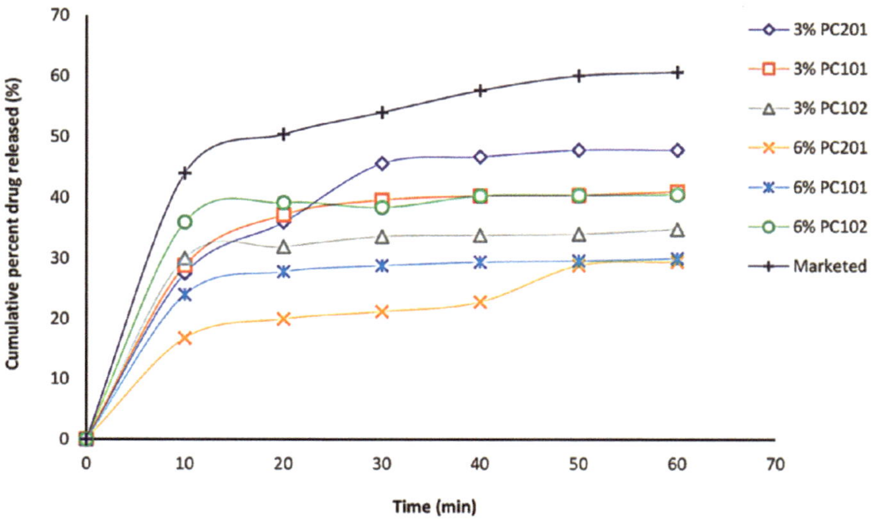

FIGURE 6.2 Dissolution plots of hydrochlorothiazide tablets containing different binders/ concentrations and that of a marketed product.

Source: Adapted from Ref. [68]

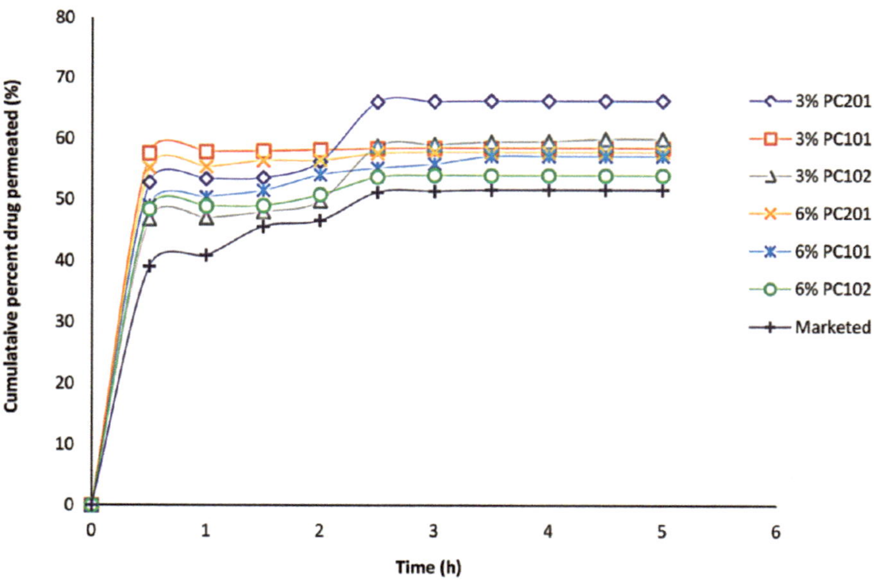

FIGURE 6.3 Permeation profile of hydrochlorothiazide tablets containing different binders/ concentrations and that of a marketed product.

Source: Adapted from Ref. [68]

6.12 EXPLORATION OF PROSOCHIT® FOR ORAL DELIVERY OF INSULIN

The ability to effectively administer insulin orally promises to be a milestone in improving adherence to insulin therapy. However, the instability of the drug in the stomach due to the inactivation by the proteolytic enzymes and the poor intestinal absorption due to the large molecular size are the major barriers to this concept [69]. The parenteral administration still remains the major route for administering this drug in diabetes management [70].

In work aimed at formulating and characterizing Prosochit®-based nanoparticles of insulin for oral delivery, drug-loaded nanoparticles were prepared by water-in-oil-in-water double emulsion – freeze drying method using the three types of Prosochit® in comparison with Prosopis gum and Callinectes chitosan as emulsifiers for the external emulsion. The formulations were subjected to particle size analysis, drug release profiling, and permeation studies [71]. The sizes of the insulin-loaded nanoparticles containing Prosochit® were found to be less than 15 nm for the inner emulsion and in the range of 128 to 160 nm for the outer emulsion. Prosochit® was also observed to limit insulin release at the gastric pH to about 10% while the values of cumulative percent drug permeated at the intestinal pH were found to be 49.8%, 53.8%, 49.1%, 54.9%, and 57.6% for the nanoparticles containing Prosopis gum, Prosochit® 201, Prosochit® 101, Prosochit® 102 and chitosan, respectively. The work showed that the three types of Prosochit®-based nanoparticulate formulations have comparable controlled release and permeation properties and are superior to the nanoparticles made of Prosopis gum or chitosan [71]. The observed performance of the nanoparticulate formulation of insulin is in consonance with the reports of McCourt et al. [73] and Yin et al. [63]. According to McCourt et al., encapsulation of insulin in a lipid system protects the drug against the proteolytic enzymes, while according to Yin et al., the presence of chitosan in insulin formulation for oral delivery enhances drug absorption.

In another work aimed at assessing the antidiabetic activity of orally administered Prosochit®-based nanoparticles of insulin in rats, the five formulations described above were assessed in comparison with orally administered purified water as a negative control and subcutaneous Actrapid® as the positive control [73]. Different groups of Wistar rats were administered daily for 14 days with different formulations. Blood sugar level was monitored and determined over 24 h, and fasting blood sugar was also taken at intervals through a period of 14 days. This was followed by plotting a graph of the percent blood glucose level against time (Figure 6.4).

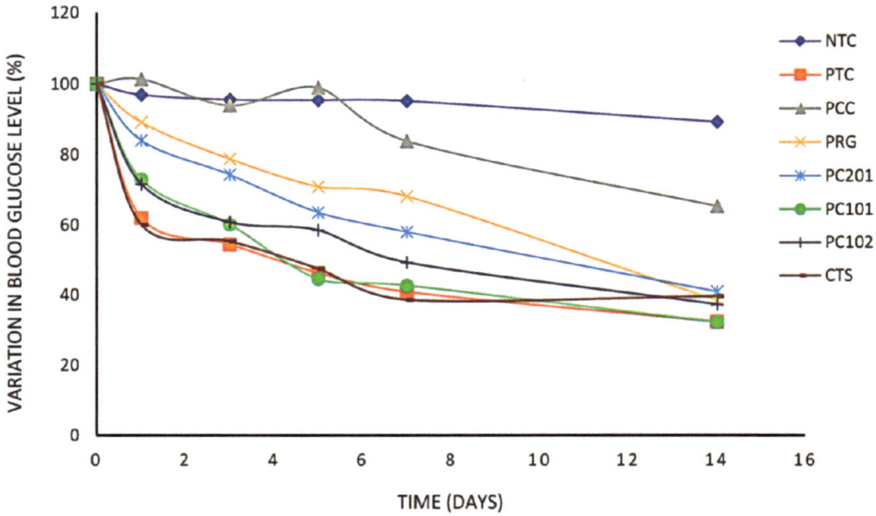

FIGURE 6.4 Antidiabetic activity of various formulations.

Note: NTC = negative control, PTC = positive control, PCC = unloaded nanoparticles, PRG = prosopis gum-based insulin nanoparticles, PC201 = Prosochit® 201-based insulin nanoparticles, PC101 = Prosochit® 101-based insulin nanoparticles, PC102 = Prosochit® 102-based insulin nanoparticles, CTS = chitosan-based insulin nanoparticles.

Source: Adapted from Ref. [73]

It was observed that there was a significant difference in the blood glucose-lowering activity of all insulin-loaded nanoparticles compared with the negative control. However, no significant difference was observed in the antidiabetic effects of the insulin-loaded nanoparticles of Prosopis gum, Prosochit® 201, Prosochit® 102, and chitosan. The insulin-loaded nanoparticles of Prosochit® 101 manifested the highest blood glucose-lowering effect. It was concluded that all the test Prosochit®-based formulations possess substantial antidiabetic activity, with the activity of the Prosochit® 101-based formulation being commensurate with that of the subcutaneous insulin [73].

The antidiabetic activity of the test formulations containing Prosochit® had been correlated with the various desirable characteristics that were built into the nanoparticles. These properties include [73]:

i. antidiabetic property of Prosopis gum itself [17];
ii. controlled-release property of Prosopis gum [74];
iii. permeation-enhancing activity of chitosan [20, 75];
iv. bioadhesive property of Prosopis gum [17];
v. co-processing of the two polymers to harness their functionalities [57];

vi. application of nanotechnology to effect particle size reduction for the purpose of enhancement of insulin absorption [71];
vii. protecting insulin from proteolytic enzymes by incorporation in the oil component of the water-in-oil-in-water multiple emulsion [72];
viii. ability of Prosochit® to promote drug delivery [64];
ix. protein-stabilizing effect of chitosan [76]; and the suitability of a chitosan-based system for the delivery of peptides and proteins, including insulin [77, 78];
x. the possibility of intact protein being absorbed [79].

Prosochit® 101 was derived from equal proportions of the two excipients. Hence, the desirable properties of the two-component polymers were optimized to enable the exceptional performance of its formulation in the delivery of insulin. The observed antidiabetic activity of Prosochit®-based nanoformulations, especially that of Prosochit® 101, being commensurate with that of the subcutaneous insulin in experimental Wistar rats needs to be further investigated in larger (higher) animals since the level of proteolytic enzymes has been found to be less in lower animals [80].

6.13 CONCLUSION

Prosopis gum is characterized by many desirable properties for drug delivery. The polymer has been applied in the delivery of many drugs. Similarly, chitosan possesses many desirable qualities and has been employed in the delivery of many drugs. There is no adverse interaction between these two polymers during co-processing, and the derived excipients, such as Prosochit® 201, 101, and 102, are endowed with the desirable properties of the two-component polymers. Prosochit® is superior to the two components as it combines the strong binding effect of Prosopis gum and the permeation-enhancing effect of Callinectes chitosan. With Prosochit®, there is synergism of characters necessary for tableting and improved character for modified drug delivery. Prosochit® has been successfully utilized for enhanced oral delivery of artemether, a drug whose oral bioavailability is limited by dissolution rate. It has also been successfully utilized for oral delivery of hydrochlorothiazide, a drug whose bioavailability is limited by both dissolution rate and absorption rate. Lately, it has been investigated for oral delivery of insulin, which has hitherto been very problematic for oral delivery. In short, Prosochit® is a novel co-processed excipient with a lot of potentials for delivering drugs that are characterized by oral delivery challenges.

KEYWORDS

- **chitosan**
- **coprocessing**
- **hydroxypropyl cellulose**
- **drug delivery**
- **Prosochit®**
- **prosopis gum**

REFERENCES

1. Naka, K. (2014). Monomers, oligomers, polymers, and macromolecules (overview). In *Encyclopedia of Polymeric Nanomaterials* (pp. 1–6).
2. Ravve, A. (2012). *Principles of Polymer Chemistry* (3rd ed.). Springer-Verlag.
3. Sung, Y. K., & Kim, S. W. (2020). Recent advances in polymeric drug delivery systems. *Biomaterials Research, 24*, 12. https://doi.org/10.1186/s40824-020-00190-7.
4. Nyamweya, N. N. (2021). Applications of polymer blends in drug delivery. *Future Journal of Pharmaceutical Sciences, 7*, 18. https://doi.org/10.1186/s43094-020-00167-2.
5. Alves, T. F. R., Morsink, M., Batain, F., Chaud, M. V., Almeida, T., Fernandes, D. A., da Silva, C. F., Souto, E. B., & Severino, P. (2020). Applications of natural, semi-synthetic, and synthetic polymers in cosmetic formulations. *Cosmetics, 7*(4), 75. https://doi.org/10.3390/cosmetics7040075.
6. Olorunsola, E. O. (2017). *Prosochit: A Group of Multifunctional Pharma Excipient*. Patent Registry. NG/P/2016/00355.
7. Moreton, R. C. (2004). Excipient functionality. *Pharmaceutical Technology, 98*, 99–119.
8. Goswami, S., & Naik, S. (2014). Natural gum and its pharmaceutical applications. *Journal of Scientific Innovative Research, 3*(1), 112–121.
9. Ahuja, R. K., Dhari, J., Goel, A., Kumar, V., & Sharma, R. (2015). Co-processing of excipients: A review on excipient development for fast dissolving tablets. *International Journal of Pharmaceutical Proff Research, 6*(3), 1264–1274.
10. Ofori-Kwakye, K., Asantewaa, Y., & Kipo, S. L. (2010). Physicochemical and binding properties of cashew tree gum in metronidazole tablet formulation. *International Journal of Pharmaceutical and Pharmaceutical Sciences, 2*(4), 105–109.
11. Froelich, A., Jakubowska, E., Jadach, B., Gadzinski, P., & Osmałek, T. (2023). Natural gums in drug-loaded micro- and nanogels. *Pharmaceutics, 15*, 759. https://doi.org/10.3390/pharmaceutics15030759.
12. Ogaji, I. S., Nep, E. I., & Audu-Peters, J. D. (2012). Advances in natural polymers as excipients. *Pharmaceutical Analysis Acta, 3*(146), 1–28.
13. Olorunsola, E. O., Bhatia, P. G., Tytler, B. A., & Adikwu, M. U. (2014). Some chemical characteristics of novel hemi cellulosic gums from seeds of Afzelia africana and Prosopis africana. *West African Journal of Pharmacy, 25*(2), 31–40.

14. Campo, V. L., Kawano, D. F., da Silva, D. B., & Carvalho, I. (2009). Carrageenans: Biological properties, chemical modifications and structural analysis–A review. *Carbohydrate Polymers, 77*, 167–180.
15. Zia, K. M., Tabasum, S., Nasif, M., Sultan, N., Aslam, N., Noreen, A., & Zuber, M. (2017). A review on synthesis, properties and applications of natural polymer-based carrageenan blends and composites. *International Journal of Biological Macromolecules, 96*, 282–301.
16. Pacheco-Quito, E. M., Ruiz-Caro, R., & Veiga, M. D. (2020). Carrageenan: Drug delivery systems and other biomedical applications. *Marine Drugs, 18*(11), 583.
17. Adikwu, M. U., Yoshikwa, Y., & Kanji, T. (2003). Bioadhesive delivery of metformin using prosopis gum with antidiabetic potential. *Biological and Pharmaceutical Bulletin, 26*(5), 662–666.
18. Majekodunmi, S. O., Olorunsola, E. O., & Uzoaganobi, C. C. (2017). Comparative physicochemical characterization of chitosan from shells of two bivalved mollusks from two different continents. *American Journal of Polymer Science, 7*(1), 15–22.
19. Olorunsola, E. O., Olayemi, O. J., Majekodunmi, S. O., & Etukudo, U. B. (2015). Extraction and physicochemical characterization of a potential multifunctional pharma-excipient from crab shell wastes. *African Journal of Biotechnology, 14*(40), 2856–2861.
20. Olorunsola, E. O., Uwah, T. O., Olayemi, O. J., & Etukudo, U. B. (2016). *Ex-vivo* evaluation of crab shell chitosan as absorption enhancer in ciprofloxacin tablet formulation. *African Journal of Biotechnology, 15*(36), 1930–1935.
21. Sonia, T. A., & Sharma, C. P. (2011). Chitosan and its derivatives for drug delivery. *Advances in Polymer Science, 243*, 23–54.
22. Ibezim, E. C., Khanna, M., Singh, S., & Uzuegbunam, C. E. (2016). Afzelia africana seed gum: Potential binder for tablet formulations. *Journal of Phytomedicine and Therapeutics, 11*, 38–48.
23. Olorunsola, E. O., Akpan, G. A., & Adikwu, M. U. (2017). Evaluation of chitosan-microcrystalline cellulose blends as direct compression excipients. *Journal of Drug Delivery, Article 8563858*, 1–8. https://doi.org/10.1155/2017/8563858.
24. Pharma Education. (2023). Tablet binders: Types and examples with concentrations. Retrieved from: https://pharmaeducation.net (accessed on 17 September 2024).
25. Olorunsola, E. O., Isah, A. B., & Zaman, Y. E. (2013). Evaluation of Borassus aethiopum starch as a binder in chloroquine tablet formulations. *International Journal of Pharmaceutical Research and Development, 5*(7), 64–70.
26. Ritthidej, G. C., Chomto, P., Pummangura, S., & Menasveta, P. (1994). Chitin and chitosan as disintegrants in paracetamol tablets. *Drug Development and Industrial Pharmacy, 20*, 2109–2134.
27. Olorunsola, E. O., Adedokun, M. O., & Akpabio, E. I. (2017). Evaluation of Callinectes chitosan as a superdisintegrant in metronidazole tablets. *International Journal of Pharmaceutical and Pharmaceutical Sciences, 9*(10), 111–118.
28. Esezobo, S., Zubair, S., & Pilpel, N. (1989). Effect of tapioca obtained from cassava (*Manihot utilissima*) on the disintegration and dissolution rate of paracetamol tablets. *Journal of Pharmacy and Pharmacology, 41*(1), 7–10.
29. Arshad, M. S., Sedhain, K., Hussain, A., Abbas, N., Mudassir, J., Mehmood, F., Irfan, M., & Latif, S. (2019). Quantification of carbon dioxide released from effervescent granules as a predictor of formulation quality using modified Chittick apparatus. *Tropical Journal of Pharmaceutical Research, 18*(3), 449–458.

30. Lodhi, V. D., Jadon, A. S., Sen, J., Jain, P. K., Thakur, B. S., Khare, B., & Jain, A. (2022). Effervescent tablets: Everything you need to know. *Asian Journal of Dental Health Sciences, 2*(4), 1–8.
31. Olorunsola, E. O., & Majekodunmi, S. O. (2017). Development of extended-release formulation of domperidone using Raphia hookeri gum as a matrix former. *Tropical Journal of Pharmaceutical Research, 16*(10), 2341–2347.
32. Olorunsola, E. O., Udoh, I. E., Majekodunmi, S. O., Odiong, I. J., & Ebong, U. O. (2021). Inhibition of gastric degradation of omeprazole using a pH-sensitive polymer as a binder in tablet formulation. *International Journal of Applied Pharmaceutics, 13*(5), 331–335.
33. Olorunsola, E. O., Adedokun, M. O., Olisakwe, P. C., & Alozie, M. F. (2022). *In vitro* evaluation of Prosopis gum matrix for duodenum-targeted delivery of omeprazole. *Journal of Applied Pharmaceutical Science, 12*(06), 164–171.
34. Avadi, M. R., Hedayatzadeh, M., & Naeimi, M. S. (2017). Preparation of alprazolam extended-release tablets and *in vitro* characterization. *Journal of Pharmaceutical Health Sciences, 5*(2), 161–170.
35. Zheng, W., Sauer, D., & McGinity, J. W. (2025). Influence of hydroxyethylcellulose on the drug release properties of theophylline pellets coated with Eudragit RS 30 D. *European Journal of Pharmaceutics and Biopharmaceutics, 59*, 147–154.
36. Tirpude, R. N., & Puranik, P. K. (2011). Rabeprazole sodium delayed-release multiparticulates: Effect of enteric coating layers on product performance. *Journal of Advanced Pharmaceutical Technology & Research, 2*(3), 184–191.
37. Guo, J., Skinner, G. W., Harcum, W. W., & Barnum, P. E. (1988). Pharmaceutical applications of naturally occurring water-soluble polymers. *Pharmaceutical Science & Technology Today, 1*, 254–261.
38. Siepmann, J., Siepmann, F., Walther, M., MacRae, R. J., & Bodmeier, R. (2008). Polymer blends for controlled release coatings. *Journal of Controlled Release, 125*, 1–15.
39. Nyamweya, N., Hoag, S. W., & Mehta, K. A. (2001). Film coating with aqueous latex dispersions: General considerations for formulating with pigments. *Pharmaceutical Technology, S4*, 8–26.
40. Kumar, R., Patil, M. B., Patil, S. R., & Paschapur, M. S. (2009). Evaluation of Anacardium occidentale gum as a gelling agent in aceclofenac gel. *International Journal of Pharmaceutical Technology Research, 1*(3), 695–704.
41. Olorunsola, E. O., & John, U. E. (2016). Solubilizing properties of hydrophilic colloid from Anacardium occidentale L. exudates in paracetamol liquid preparations. *African Journal of Pharmaceutical Research & Development, 8*(2), 147–152.
42. Olorunsola, E. O., & Majekodunmi, S. O. (2016). Emulsifying properties of Afzelia gum in liquid paraffin emulsion. *International Journal of Pharmacy and Pharmaceutical Sciences, 8*(11), 195–198.
43. Adedokun, M. O., Olorunsola, E. O., & Anselem, E. L. (2017). Comparative evaluation of metronidazole suspensions formulated with Raphia africana hydrocolloid and other natural polymers. *Journal of Applied Pharmaceutical Science, 7*(6), 121–126.
44. Huang, C. C., Du, M. X., Zhang, B. Q., & Liu, C. Y. (2022). Glass transition temperatures of copolymers: Molecular origins of deviation from the linear relation. *Macromolecules, 55*(8), 3189–3200.
45. Kesharwani, P., & Greish, K. (2022). *Polymeric Micelles for Drug Delivery*. Woodhead Publishing. pp. 675–684.

46. Olayemi, O. J., & Nock-Anyebe, S. (2021). Evaluation of novel co-processed excipient for fast disintegration of aspirin tablet formulation. *Journal of Pharmaceutical Bioresources, 18*(1), 1–11.
47. Ngwuluka, N. C., Nep, E. I., Ochekpe, N. A., Odumosu, P. O., & Olorunfemi, P. O. (2015). Eudragit 100 and polysaccharide polymer blends as matrices for modified-release drug delivery II: Swelling and release studies. *Tropical Journal of Pharmaceutical Research, 14*(12), 2163–2170.
48. Marwaha, M., Sandhu, D., & Marwaha, R. K. (2010). Co-processing of excipients: A review of excipient development for improved tableting performance. *International Journal of Applied Pharmaceutics, 2*(3), 41–47.
49. Russell, R. (2004). Synthetic excipients challenge all natural organics—Offer advantages/challenges to developers and formulators. *Pharmaceutical Technology, 27*, 38–50.
50. Agboola, D. A. (2004). Prosopis africana (Mimosaceae): Stem, roots, and seeds in the economy of the savannah areas of Nigeria. *Economic Botany, 58*(suppl), 34–42.
51. Olorunsola, E. O., Bhatia, P. G., Tytler, B. A., & Adikwu, M. U. (2015). Physicosurface properties of Afzelia and Prosopis hemi cellulosic gums: Potential surface-active agents. *IOSR Journal of Pharmacy and Biological Sciences, 10*(4, Version 3), 1–7.
52. Olorunsola, E. O., Bhatia, P. G., Tytler, B. A., Adedokun, M. O., & Adikwu, M. U. (2017). Dissolution and permeation characteristics of artemether tablets formulated with two gums of different surface activity. *Tropical Journal of Pharmaceutical Research, 16*(5), 981–988.
53. Olorunsola, E. O., Adedokun, M. O., Bhatia, P. G., Tytler, B. A., & Adikwu, M. U. (2017). Optimization of artesunate delivery by formulation in a delayed-release Prosopis hemicellulose matrix. *Journal of Applied Pharmaceutical Science, 7*(5), 142–146.
54. Akin-Oriola, G., Anetekhai, M., & Olowonirejuaro, K. (2005). Morphometric and meristic studies in two crabs: *Cardiosoma armatum* and *Callinectes pallidus*. *Turkish Journal of Fisheries and Aquatic Sciences, 5*, 85–89.
55. Burrow, F., Louime, C., Abazinge, M., & Onokpise, O. (2007). Extraction and evaluation of chitosan from crab exoskeleton as a seed fungicide and plant growth enhancer. *American-Eurasian Journal of Agricultural and Environmental Sciences, 2*(2), 103–111.
56. Chiappisi, L., & Gradzielski, M. (2015). Co-assembly in chitosan-surfactant mixtures: Thermodynamics, structures, interfacial properties, and applications. *Advances in Colloid and Interface Science, 220*, 92–107.
57. Olorunsola, E. O., & Usungurua, S. G. (2018). Effect of polymer ratio on the quality of Prosochit®—A co-processed excipient of Prosopis gum and crab shell chitosan. *Tropical Journal of Pharmaceutical Research, 17*(9), 1693–1699.
58. Iqbal, M. S., Massey, S., Akbar, J., Ashraf, C. M., & Masih, R. (2013). Thermal analysis of some natural polysaccharide materials by isoconversional method. *Food Chemistry, 140*(1–2), 178–182.
59. Horvat, M., Mestrovic, E., Danilovski, D., & Craig, D. Q. M. (2005). An investigation into the thermal behavior of a model drug mixture with amorphous trehalose. *International Journal of Pharmaceutics, 294*(1–2), 1–10.
60. Aulton, M. E. (2018). Powder flow. In M. E. Aulton & K. M. G. Taylor (Eds.), *The Design and Manufacture of Medicines* (pp. 189–200). Churchill Livingstone Elsevier.

61. Sun, X. F., Wang, H. H., Jing, Z. X., & Mohanathas, R. (2013). Hemicellulose-based pH-sensitive and biodegradable hydrogel for controlled drug delivery. *Carbohydrate Polymers, 92*(2), 1357–1366.
62. Attama, A. A., Adikwu, M. U., & Okoli, N. (2000). Studies in bioadhesive granules 1: Granules formulated with Prosopis africana gum. *Chemical & Pharmaceutical Bulletin, 48*(5), 734–737.
63. Yin, L., Ding, J., He, C., Cui, L., Tang, C., & Yin, C. (2009). Drug permeability and mucoadhesive properties of thiolated chitosan nanoparticles in oral insulin delivery. *Biomaterials, 30*, 5691–5700.
64. Olorunsola, E. O., Tologbonse, A. A., Onwuka, N. A., & Adikwu, M. U. (2019). Enhanced oral delivery of artemether: An application of Prosochit®. *Journal of Applied Pharmaceutical Science, 9*(04), 133–136.
65. Olorunsola, E. O., Bhatia, P. G., Tytler, B. A., & Adikwu, M. U. (2016). Compatibility study of cashew and Prosopis gums with some artemisinin derivatives. *Nigerian Journal of Pharmaceutical Sciences, 50*(2), 35–53.
66. Sinha, V. R., Singla, A. K., Wadhawan, S., Kaushik, D., Kumria, R., Bansal, K., & Dhawan, S. (2004). Chitosan microspheres as a potential carrier for drugs. *International Journal of Pharmaceutics, 274*(1–2), 1–33.
67. Samphui, P., Devi, V. K., Clara, D., Malviya, N., Ganguly, S., & Desiraju, G. R. (2015). Cocrystals of hydrochlorothiazide: Solubility and diffusion/permeability enhancement through drug-coformer interactions. *Molecular Pharmaceutics, 12*(5), 1615–1622.
68. Olorunsola, E. O., Adedokun, M. O., Samuel, U. O., & Joseph, M. G. (2021). Effects of Prosochit® binder on the dissolution and permeation of a BCS class IV drug. *Tropical Journal of Natural Product Research, 5*(6), 1138–1141.
69. Olorunsola, E. O., Alozie, M. F., Davies, K. G., & Adedokun, M. O. (2021). Advances in the science and technology of insulin delivery: A review. *Journal of Applied Pharmaceutical Science, 11*(8), 189–196.
70. Olorunsola, E. O., Udoh, I. E., Ekott, M. B., Alozie, M. F., & Davies, K. G. (2023). Biopharmaceutics and clinical outcomes of emerging dosage forms of insulin: A systematic review. *Diabetes Epidemic & Management, 9*, 100120.
71. Olorunsola, E. O., Davies, K. G., Ibiang, K. P., Esukpa, P. C., Uwaechi, E. G., & Ahsan, F. (2023). Prosochit®-based nanoparticulate system of insulin for oral delivery: Design, formulation, and characterization. *Journal of Applied Pharmaceutical Science, 13*(03), 044–052.
72. McCourt, M., Mielnicki, L., Hughes, J., Irving, M. Q., Schentag, J., Ghanim, H., & Dandona, P. (2020). 1013-P: Modification of inflammatory and Alzheimer markers with intracellular delivery of insulin using cholestosomes. *Diabetes, 69*(Suppl 1). https://doi.org/10.2337/db20-1013-P.
73. Olorunsola, E. O., Davies, K. G., Essien, E. B., Alozie, M. F., Adedokun, M. O., & Ahsan, F. (2022). Orally administered Prosochit®-based nanoparticles of insulin ameliorate alloxan-induced diabetes in rats. *Scientific Reports, 90*, 66. https://doi.org/10.3390/scipharm90040066.
74. Henciya, S., Seturaman, P., James, A., Tsai, Y., Nikam, R., Wu, Y., Dahms, H., & Chang, F. (2017). Biopharmaceutical potentials of Prosopis spp. (Mimosaceae, Leguminosa). *Journal of Food and Drug Analysis, 25*, 187–196. https://doi.org/10.1016/j.jfda.2017.02.001.

75. Majekodunmi, S. O., Olorunsola, E. O., Ofiwe, U. C., Udobre, A. S., & Akpan, E. (2017). Material properties of chitosan from shells of *Egeria radiata*: Drug delivery considerations. *Journal of Coastal Life Medicine, 5*(7), 321–324.
76. Mohammed, M. A., Syeda, J. T. M., Wasan, K. M., & Wasan, E. K. (2017). An overview of chitosan nanoparticles and its application in non-parenteral drug delivery *Pharmaceutics, 9*(4), 53.
77. Amidi, M. E., Mastrobattista, E., Jiskoot, W., & Hennink, W. E. (2010). Chitosan-based delivery systems for protein therapeutics and antigens. *Advanced Drug Delivery Reviews, 62*, 59–82.
78. Erel, G., Kotmakc, I. M., & Akbaba, H. (2016). Nano encapsulated chitosan nanoparticles in emulsion-based oral delivery system: *In-vitro* and *in-vivo* evaluation of insulin-loaded formulation. *Journal of Drug Delivery Science and Technology, 36*, 161–167.
79. Gardner, M. L. (1988). Gastrointestinal absorption of intact proteins. *Annual Review of Nutrition, 8*, 329–350.
80. Gorrill, A. D. L., Schingoethe, D. J., & Thomas, J. W. (1968). Proteolytic activity and *in vitro* enzyme stability in small intestinal contents from ruminants and non-ruminants at different ages. *Journal of Nutrition, 96*(3), 342–348.

CHAPTER 7

Advances in Controlled Drug Delivery Using Grafted Copolymer

S. GIRIDHAR REDDY[1] and H. C. ANANDA MURTHY[2]

[1]*Department of Chemistry, Amrita School of Engineering, Amrita Vishwa Vidyapeetham, Bangalore, India*

[2]*Department of Applied Chemistry, School of Natural Science, Adama Science and Technology University, Adama, Ethiopia*

ABSTRACT

Grafted copolymers have become a reliable approach in drug delivery, offering the controlled release of drugs that can enhance their effectiveness while minimizing potential toxicity. This approach involves attaching a hydrophilic polymer to a hydrophobic drug molecule, resulting in a stable drug-polymer conjugate. Grafted copolymers offer numerous benefits, including improved drug solubility, enhanced stability, sustained release, and targeted delivery. Recent advancements in this field have focused on developing innovative grafting techniques, optimizing polymer structures, and enhancing drug loading efficiency. This chapter comprehensively overviews the latest advancements in CDD utilizing grafted copolymers. It covers various topics, including the types of polymers utilized, drug release mechanisms, and various drug delivery applications. The chapter also addresses the challenges and prospects in this area, emphasizing the importance of improving the biocompatibility and biodegradability of grafted copolymers for safe clinical utilization.

Graft Copolymers of Polysaccharides for Drug Delivery Systems.
Rishabha Malviya and Sonali Sundram (Eds.)

7.1 INTRODUCTION

Advances in CDD using grafted copolymer" is a chapter exploring recent advancements in drug delivery techniques involving grafted copolymers. Drug delivery is a critical field of study with significant implications for effective disease treatment. Grafted copolymers have garnered significant interest as a favorable process for achieving CDD owing to their unique attributes, such as exceptional biocompatibility, high capacity for drug loading, and the capability to release drugs in a controlled and regulated manner [1].

The chapter starts with an introduction to the drug delivery concept and its associated challenges. It then provides an overview of grafted copolymers and their properties, including the grafting techniques used to synthesize them. The chapter then discusses the various applications of grafted copolymers in drug delivery, including their use in cancer therapy, the treatment of minimally invasive devices, ophthalmology, scaffolds, polymeric micelles, and 3D/4D printing.

The chapter also explores the latest breakthroughs in drug delivery utilizing grafted copolymers, including the progress in stimuli-responsive copolymers capable of drug release triggered by specific stimuli and the application of nanotechnology for targeted drug delivery [2, 3].

The chapter is intended for researchers and students in drug delivery, polymer science, and biomedical engineering.

CDD provides a precise and predetermined rate of drug release over some time. CDD aims to optimize the drug's therapeutic effect while minimizing any side effects or toxicity [4].

In a CDD system, the drug is typically encapsulated or conjugated with a delivery vehicle, such as a polymer or lipid-based carrier, that controls the release of the drug. Various techniques, such as diffusion, degradation, or osmosis, can control the drug release. The drug release rate can be tailored to suit the specific therapeutic needs of the patient.

CDD has several advantages over conventional drug delivery, including improved patient compliance, reduced dosing frequency, and enhanced therapeutic efficacy. CDD systems can also minimize the risk of toxicity and side effects by delivering the drug to the targeted place, thereby reducing the exposure of non-target tissues [5–7].

7.2 GRAFT COPOLYMER IN CDD

Graft copolymers have played a significant role in recent advancements in drug delivery. Graft copolymers are formed by attaching a functional group or side chain to a pre-existing polymer backbone, resulting in a copolymer with unique properties. Incorporating graft copolymers into drug delivery systems (DDSs) has enabled the CDD systems, improved biocompatibility, and reduced toxicity.

DDS's most commonly used graft copolymer is polyethylene glycol (PEG). PEG-based copolymers have been experimented with for their ability to increase drug circulation time and improve drug solubility. PEGylation of drugs also enhances their stability and reduces their immunogenicity.

Additional graft copolymers employed in pharmaceutical delivery encompass poly(lactic-co-glycolic acid) (PLGA) and PEO-poly(propylene oxide) block copolymers. These copolymers have created specific drug delivery mechanisms like nanostructured lipid carriers and polymeric micelles [8].

Graft copolymers have contributed significantly to recent advancements in drug delivery by improving drug stability, enhancing biocompatibility, and enabling targeted drug delivery. As research continues, graft copolymers in drug delivery will likely expand, resulting in more effective and efficient DDSs.

The applications of graft copolymers in various fields are furnished below.

7.2.1 GRAFT COPOLYMER CDD APPLICATIONS IN CANCER

The limitations of systemic chemotherapy for cancer treatment are due to its toxicity and the potential to damage healthy cells. This chapter summarizes the exploration of DDS to prevent the harmful effects of drugs on the entire body. It delves into various types of DDSs, including polymeric conjugates, nano or microparticles, hydrogels, and microneedles. The chapter also highlights recent advancements to enhance the safety, stability, and controlled release properties of these DDSs. Additionally, it addresses the drawbacks of long-acting anticancer therapies and explores potential future technological advancements in this field [9].

Various researchers conducted research under graft copolymer, and their contributions are discussed below.

7.2.1.1 GRAFTING COPOLYMER APPLICATIONS IN BREAST CANCER

Breast cancer-related deaths in women are increasing globally. Despite advancements in awareness and diagnosis that have contributed to reduced mortality rates, challenges persist in the form of late detection and drug resistance, which hinder the effectiveness of treatment and contribute to relapse and metastasis. Core-shell mesoporous silica nanoparticles (MSNPs) have garnered attention as a potential platform for drug delivery owing to their advantageous features, such as a large surface area, low density, and tunable size and morphology. MSNPs can be functionalized for site-specific drug delivery, improving pharmacokinetics (PKs) and biodistribution. Rani et al. [10] discuss the preparation, DDS, and emerging applications of MSNPs in treating breast cancer. The authors emphasize the potential of MSNPs to enhance treatment effectiveness, minimize side effects, and improve overall treatment outcomes.

Ehsanimehr et al. [11] developed nanocarriers to deliver doxorubicin (DOX) chemotherapy. They used nanocrystalline cellulose as the base material and modified it through chlorination and atom transfer radical polymerization. Two types of nanocarriers were synthesized, loaded with DOX, and characterized using various techniques. The drug release and impact on breast cancer cells (MCF7 cells) were evaluated.

In their study, Kong et al. [12] emphasize the detrimental effects of chemotherapy drug toxicity and the emergence of resistant tumor cell clones on the efficacy of treatment for highly aggressive triple-negative breast cancer (TNBC). The article examines the utilization of nanoparticle delivery systems as a comprehensive targeted therapy for TNBC. This underscores the capacity of these systems to efficiently transport and release diverse components, encompassing chemotherapeutic drugs, active gene fragments, and immune-enhancing factors simultaneously. Additionally, the synergistic effect of combining nanoparticle delivery systems with photothermal ablation therapy is explored, which has the potential to inhibit genes associated with tumor proliferation and activate the immune system. The focus is on the research advancements in nanoparticle delivery systems for comprehensive targeted therapy in TNBC, discussing their advantages, limitations, prospects, and future developments. Their research offers valuable insights into nanoparticle-DDSs' potential for treating TNBC.

7.2.2 CLAY NANOSHEETS GRAFT COPOLYMER

Huang et al. [13] aimed to enhance the stability of montmorillonite (MMT) as a DDS. They achieved this by developing PEG-CS/MMT nanosheets through a grafting process. The nanosheets had PEG-rich surfaces, ensuring excellent dispersion in serum environments. By adjusting the ratio of PEG-CS to MMT, the researchers observed a transition from multi-layered to exfoliated structures. The multi-layered nanosheets exhibited superior drug-loading capacity compared to the exfoliated ones. The DOX@PEG-CS/MMT nanosheets were taken up by TRAMP-C1 cells and released DOX gradually, leading to cell apoptosis. This study presents a novel strategy for achieving stable dispersion of MMT nano clays in drug delivery applications.

7.2.3 PH-RESPONSIVE POLYMERIC NANOCAPSULES

Massoumi et al. [14] developed pH-responsive hollow nanocapsules (HNCaps) with PEGylation using a novel method. They synthesized SiO_2 nanoparticles and modified them with MPS, followed by copolymerization with triethylene glycol dimethacrylate (TEGDMA), acrylic acid, and MPS-modified SiO_2 nanoparticles to form SiO_2@PAA core-shell nanoparticles. The SiO_2 core was then removed, resulting in PAA hollow nanocapsules. Thiol-end-capped PEG was grafted onto these nanocapsules through a thiol-Ene 'click' reaction, forming PAA-g-PEG HNCaps. The HNCaps were loaded with DOX to assess their drug loading, encapsulation efficiency, and pH-dependent drug release. The study revealed promising properties for cancer chemotherapy, including favorable characteristics and potent cytotoxicity against cancer cells.

7.2.4 PACLITAXEL AS AN ANTI-CANCER DRUG

Ramezanpour et al. [15] acknowledged the difficulties associated with poor water solubility and limited Paclitaxel (PTX) permeability in biological tissues. Traditionally, PTX has been formulated using Cremophor EL and dehydrated ethanol to address these challenges. However, the undesirable side effects of this formulation have prompted researchers to explore alternative PTX formulations based on polysaccharides. Polysaccharides are biocompatible, biodegradable, and easily chemically modified, making them an excellent alternative to Cremophor EL. The emphasis is placed on the latest progress in polysaccharide-based formulations of PTX, aiming to enhance its efficacy and mitigate the adverse effects of PTX in cancer therapy.

7.2.5 DEXTRAN-GRAFT-POLYACRYLAMIDE

Grebinyk et al. [16] aimed to develop dextran-graft-polyacrylamide (D-g-PAA) copolymers for precise drug delivery in cancer treatment. They synthesized anionic D-g-PAA nanoparticles (D-g-PAAan) that release chemotherapeutic drugs, such as DOX and Cisplatin, in response to pH changes. The D-g-PAAan nanoparticles exhibited enhanced effectiveness against lung carcinoma cells while preserving the viability of normal lung cells. Furthermore, the nanoparticles loaded with the drug D-g-PAAan exhibited enhanced drug accumulation and exerted cytotoxic, prooxidant, and proapoptotic effects on lung carcinoma cells within a 3D multicellular spheroid model. These findings highlight the capability of D-g-PAAan nanoparticles to serve as a TDD system for cancer treatment.

7.2.6 DOXORUBICIN CDD BY IRON OXIDE NANOPARTICLES

Khabibullin et al. [17] describe a study comparing the properties of four potential DOX carriers: iron oxide nanoparticles (IONs), which are functionalized with different polymers (polyethyleneimine (PEI), polystyrene sulfonate (PSS), dextran, and porous carbon). The study evaluates the loading and release capacities of doxorubicin from these carriers, as well as their toxicity and potential effects on blood clotting. The findings suggest that the particles that underwent modification with PEI exhibit the greatest loading capacity. Conversely, the particles modified with PSS demonstrate a slow drug release, which could be advantageous for achieving a prolonged inhibitory effect on tumors. The toxicity assessment using the Neuro2A cell line shows no negative effect. The study suggests that using nanoplatforms for drug delivery could be a better approach to cancer treatment, and the results obtained can help develop new drug delivery platforms.

7.2.7 NANOVEHICLES FOR TARGETED CANCER TREATMENT

Zhang et al. [18] examine the potential of pH-responsive nanovehicles for targeted cancer treatment. They discuss the acidic nature of the tumor microenvironment (TME) and how synthetic polypeptide poly(l-histidine) (PLH)-based pH-responsive nanovehicles can release chemotherapeutic agents in response to the TME's pH. PLH exhibits pH-dependent hydrophobic-to-hydrophilic phase transitions, leading to the destabilization of nanovehicles and subsequent drug release. The article explores the advanced design and construction of such nanovehicles and addresses future challenges and prospects regarding translating this biomaterial into clinical applications.

7.2.8 ANTICANCER DRUG DELIVERY BY BOTTLEBRUSH POLYMERS

Wang et al. [19] introduce a novel approach to producing precisely structured bottlebrush polymers using RAFT polymerization. The procedure entails synthesizing diazoacetate macromonomers of polystyrene and PEG with precisely regulated molecular weights. Subsequently, these macromonomers undergo polymerization using an allyl PdCl/L catalyst, forming well-designed bottlebrush polymers. An amphiphilic bottlebrush polymer is created by polymerizing diazoacetate macromonomers of PEG using a Pd(II)-terminated polystyrene bottlebrush polymer as the macroinitiator. The amphiphilic polymer can spontaneously arrange itself into a highly structured core-shell micelle configuration. This micelle demonstrates outstanding biocompatibility, indicating its potential for use in drug delivery applications. The study showcases the versatility and efficiency of bottlebrush polymers as building blocks for designing innovative materials with customized properties.

7.2.9 POLYURETHANE GRAFT ON CYCLODEXTRIN FOR TUMOR TREATMENT

Shukla et al. [20] present a novel graft copolymer architecture for drug delivery in melanoma cancer treatment. The copolymer design allows for control over the hydrophilicity of cyclodextrin while maintaining a balanced hydrophilic-hydrophobic nature. Compared to cyclodextrin, the graft copolymer exhibits enhanced thermal and mechanical stability. Sustained drug release is achieved through specific interactions between the copolymer and the drug. Cellular studies confirm the biocompatibility of the graft copolymer, while animal experiments demonstrate its effectiveness in suppressing tumors without severe side effects. The unique architecture of the graft copolymer makes it a promising drug delivery vehicle for melanoma cancer treatment.

7.2.10 STAR-SHAPED COPOLYMERS IN SOLID TUMORS

In their study, Kostka et al. [21] introduced novel biodegradable star-shaped copolymers that offer potential advancements in drug delivery applications. The copolymers were prepared by attaching linear N-(2-hydroxypropyl)methacrylamide copolymers onto a dendritic core composed of 2,2-bis(hydroxymethyl)propionic acid (bisMPA). The size of the star copolymers was precisely managed by selecting the dendritic core type, generation, molecular weight of the copolymer, and polymer-to-core ratio. The copolymers demonstrated biodegradability and outperformed the first generation with non-degradable PAMAM cores in suppressing tumor growth in mice. This study highlights the ability of these new biodegradable star copolymers to be highly effective DDS with improved tumor accumulation.

7.2.11 ANTICANCER DRUG 5-FLUOROURACIL (5-FU)

El-Zeiny et al. [22] conducted research on the synthesis and characterization of novel nanogel structures by grafting β-Cyclodextrin onto N-vinylcaprolactam, aiming to develop a carrier for the anticancer drug 5-FU. The nanogels were synthesized in aqueous media using different ratios of β-Cyclodextrin and N-vinylcaprolactam through free radical graft copolymerization. The findings revealed that nanogels prepared with a 1:5 ratio of β-Cyclodextrin to N-vinylcaprolactam exhibited the highest loading capacity and controlled release behavior for 5-FU. The loading behavior was described using Langmuir and Freundlich models, while the chemical uptake of drug molecules was analyzed using the Dubinin-Radushkevich model. The nanogels exhibited low cytotoxicity on normal cells and increased toxicity on cancer cells upon loading with 5-FU.

7.2.12 CO-DELIVERY OF ANTI-CANCER DRUGS

In their study, Bakr et al. [23] demonstrated the creation of two different types of nanocomposites by combining polylactic acid (PLA) and nanocellulose (NC). These nanocomposites were designed to serve as carriers for two anti-cancer drugs, 5-FU and curcumin. Fe_3O_4 NPs (M) were introduced into the two combinations using blending and self-assembly techniques to incorporate both drugs. The researchers examined these drug-loaded nanocarriers' loading capacity, release behavior, and effectiveness against bacteria and various cell lines. The results indicated that the micelles composed of the copolymer (M-PLA-co-NC) showed a reduced release rate and enhanced antibacterial and antitumor efficacy compared to the blend formulation (PLA/M-NC). These findings indicate that the copolymeric micelle is promising as nanomedicine for treating cancer and bacterial infections.

Demirel et al. [24] have introduced a novel approach to creating a nanoscale platform for delivering hydrophobic drugs and oligonucleotides to cancer cells. This platform combines reduced graphene oxide (rGO) with a copolymer containing pyrene groups of PEG and an azide functional group. By utilizing the azide functionality, cationic copolymers and targeting moieties such as the EPPT1 peptide, specifically binding to MUC1 in early tumorigenesis, can be attached. The resulting PAMP-CP-rGO nanoplatforms demonstrate pH-adjustable controlled drug release and efficient transfection capabilities, targeting GAPDH protein levels. *In vitro*, cytotoxicity evaluations reveal that EPPT1-targeted PAMP-CP-rGO loaded with DOX exhibits concentration-dependent toxicity against cancer cells. These findings highlight the potential of these versatile nanoplatforms for the selective and safe transportation of various cargoes to tumor sites.

Bláhová et al. [25] describe the development of amphiphilic copolymers for efficient dual drug delivery in cancer therapy. The PCL-graft-pHPMA copolymers were synthesized using controlled RAFT polymerization and click chemistry, resulting in well-designed polymers with diverse molecular weights and physical properties. The copolymers are self-assembled into stable micelles of a diameter of around 25 nm and a low critical micelle concentration (CMC). The micelles could encapsulate hydrophobic and pH-sensitive chemotherapeutic drugs, with a total drug payload of up to 17 wt%. The micelles were stable in buffers but were cleaved in lipase, indicating degradation and drug release inside cancer cells.

Qiu et al. [26] discuss the challenges in treating cancer due to its complex nature and heterogeneity and the potential of using nanotechnology-based combined therapy to enhance anticancer effects. Different cargo molecules, including nucleic acids, therapeutic drugs, and imaging agents, are assessed regarding their potential for clinical applications as medications. The publication emphasizes the benefits of utilizing nanomedicine for co-delivering small-molecule drugs and nucleic acids in anticancer treatments. It examines the obstacles that hinder the clinical implementation of such approaches and provides insights into prospects for rationally guiding combined tumor therapies.

Tao et al. [27] studied synthesizing and characterizing a dual-sensitive amphiphilic graft polyurethane, FA-PUSS-gimi-mPEG, designed to release anticancer drugs to tumor cells efficiently. The self-assembled micelles generated in a water-based solution possess a negative charge and maintain stability under typical physiological circumstances. In mildly acidic conditions, the PEG layer experiences dissociation caused by the cleavage of the benzoic-imine bond. Consequently, the micelles' surface charge transforms, and folic acid is exposed on the surface. The FA-PUSS-gimi-mPEG micelles exhibit effective loading of DOX, which is rapidly released in acidic conditions and in the presence of glutathione (GSH). These micelles possess dual-sensitive properties that enhance cellular uptake and promote drug release within tumor cells. This capability facilitates the efficient delivery of anticancer drugs and effectively inhibits cellular proliferation.

Movagharnezhad et al. [28] presented a study where magnetite nanoparticles (MNP, Fe_3O_4) were surface-modified using a cellulose polymer. To achieve this, cellulose was changed into its bromoacetylated derivative (BACell), which was then used to functionalize the MNPs. The researchers developed an atom transfer radical polymerization (ATRP) method to covalently immobilize N-vinylpyrrolidone (NVP) on the Fe_3O_4/BACell surface, resulting in the production of Fe_3O_4/Cellulose-grafted NVP (Fe_3O_4/Cell-NVP). The study also investigated the loading capacity and released DOX profiles, a model drug, from Fe_3O_4/Cell-NVP. The results demonstrated that the delivery of DOX from the loaded nanoparticles was well-controlled during the release period. Based on these findings, Fe_3O_4/Cell-NVP is suggested as a suitable candidate for the CDD and targeted delivery of cancer treatment.

7.2.13 POLYURETHANE-GRAFTED DEXTRIN IN CANCEROUS HELA

Shukla et al. [29] have presented a study detailing the synthesis of a new copolymer consisting of polyurethane grafted onto Dextrin. The degree of grafting was determined using NMR, and the observed morphology of the copolymer exhibited strip-like structures. In laboratory studies, it was observed that the release of the drug exhibited a prolonged pattern, and analyses using spectroscopic techniques confirmed evident interactions between the polymer and the drug. The biocompatibility of the copolymers was evaluated through cellular studies utilizing cancerous HeLa cells. The MTT assay was employed to assess cell viability and cell adhesion. The cytotoxicity of the graft copolymers was demonstrated, leading to a notable decrease in cell survival.

7.2.14 SUPRAMOLECULAR NANOGELS

Maryam Monajati et al. [30] describe the development of a biocompatible nanogel containing L-asparaginase, a therapeutic agent used to treat leukemia, to improve the enzyme's stability. The nanogel was made by combining ASNase-conjugated copolymers with crosslinked α-cyclodextrin to form the inclusion of complex nanogels. The resulting nanogels were homogenous and spherical, with a diameter of 148 nm. The immobilized ASNase showed a 1.6-fold increase in affinity for asparagine substrate and exhibited higher stability than the free enzyme. This immobilized ASNase-bisCD nanogel is a promising biomedical application for cancer treatment.

7.3 POLYMERIC MICELLES IN CDD

Polymeric micelles are nanocarriers designed to deliver active pharmaceutical ingredients (APIs) with specific characteristics such as low bioavailability, biodegradability, and toxicity. Different polymers like diblock, graft, and triblock fabricate specialized micelles with distinctive characteristics like targeted drug delivery and controlled release. Polymeric micelles demonstrate significant potential as advanced nanocarriers for drug delivery, offering enhanced performance in complex intercellular and intracellular environments [31].

Polymeric micelles are engineered to enhance the solubility of drugs with limited water solubility and safeguard molecules sensitive to water. They are designed to prolong circulation in the body, decrease drug toxicity, and enable precise targeting, particularly for tumors. Different stimuli, including pH levels, redox reactions, glucose concentrations, reactive oxygen species, hypoxia, and enzymes, can activate these micelles. Recent progress has facilitated the creation of polymeric micelles that integrate nucleic acids, phytoconstituents, surface modifications, ligand conjugation, protein-based components, and biosensors. Polymeric micelles have been extensively studied as potential theranostic agents to treat and diagnose various conditions, including cancer, arthritis, and diabetes [32].

7.3.1 POLYMERIC MICELLES IN CANCER THERAPIES

Polymeric micelles can regulate drug release patterns and enhance drug concentration at specific locations, making them highly suitable for medications with poor solubility. They demonstrate biocompatibility and have minimal toxicity towards healthy cells, positioning them as a promising option for cancer treatment. These micelles can accumulate in the TME and can be targeted to specific cancer sites using various stimuli. Polymeric micelles have significantly advanced drug and gene delivery in cancer treatment, including applications such as crossing the blood-brain barrier, gene therapy, photothermal therapy, and immunotherapy. Some polymeric micelles have already been approved for therapeutic use, while others are undergoing clinical trials for various cancer therapies [33].

Mehan et al. [34] discuss the utilization of polymeric micelles as carriers for drug delivery in nano-sized systems, with a specific focus on treating skin cancer. These micelles are constructed using synthetic block copolymers and graft copolymers, which can be combined with bilayer lipid membranes and surfactants. Surfactant micelles are notable for their stability and ability to encapsulate drugs with poor water solubility or amphiphilic.

Adeli et al. [35] present a study on developing novel dual-responsive amphiphilic micelles using a host-guest interaction between PHEMA-g-(PCL-BM) and β-CD-star-PMAA-b-PNIPAM. These micelles exhibit responsiveness to changes in temperature and pH, and they were successfully loaded with DOX with high encapsulation efficiency. The micelles demonstrate self-assembly capability, low CMC value, and efficient drug release. The biocompatibility of the micelles was confirmed using the human breast cancer cell line MCF-7, and the DOX-loaded micelles exhibited remarkable anti-cancer activity.

Polymeric micelles are self-assembled nanostructures formed by amphiphilic macromolecules, offering unique properties and benefits. They improve drug solubility, enhance bioavailability, and enable targeted delivery of drugs. Unde et al. [36] explore using polymeric micelles to deliver poorly soluble phytoconstituents in herbal formulations. Their research incorporates phytoconstituents like Artemisinin, Curcumin, Kava-kava, and Berberine into polymeric micelles to overcome low bioavailability and achieve diverse therapeutic effects. Polymeric micelles outperform other nanocarriers by exhibiting high entrapment efficiency, loading capacity, and targeted accumulation at specific sites.

Niesyto et al. [37] synthesized graft copolymers using a choline ionic liquid and fusidate anions as pharmaceutical anions through atom transfer radical polymerization. The resulting copolymers and their conjugates containing the antibiotic Rifampicin demonstrated the formation of micellar structures capable of encapsulating significant drug quantities. The dual-drug systems exhibited controlled release of both drugs, albeit at slightly lower rates than single-drug systems. These findings highlight the promising capability of these systems as delivery platforms for combined therapy targeting drug-resistant strains. This approach offers the advantage of delivering two drugs simultaneously within a single formulation, eliminating the need for separate administration.

Kuperkar et al. [38] present a summary of the latest progress in the self-assembly of amphiphilic block copolymers into nanoscale periodic structures, showcasing their potential for applications in the fabrication of nanomaterials. The article discusses various block copolymers and their ability to form microdomains and undergo mineralization in selective solvents. Different polymer micelles and polymersomes are explored, including DHBCs, shell and core-cross-linked structures, core-shell-corona configurations, schizophrenic structures, crew cut architectures, and Janus structures. The article also examines PISA and CDSA techniques for self-assembly.

Singam et al. [39] discuss the utilization of natural polymers as preferred over synthetic polymers for DDS. The researchers synthesized a micellar DDS by grafting PEG onto ethyl cellulose. Subsequently, they characterized its properties, cell toxicity, hemocompatibility, and drug release kinetics using DOX. The study reveals that the micelles exhibited significant efficacy in eliminating cancer cells, specifically MDA-MB-231 and MCF-7, while displaying non-toxicity towards normal cells. This indicates the potential of these micelles as a controlled DDS for cancer treatment. The authors propose conducting further *in vivo* studies to explore the full capabilities of this material.

Zeng et al. [40] studied drug-loaded unimolecular micelles formed by PAMAM(G5)-PCBMA and PAMAM(G5)-PEGMA for drug delivery applications. Using dissipative particle dynamics simulations, they found that PAMAM(G5)-PCBMA micelles possess a stable core-shell structure, with the hydrophobic core acting as a reservoir for DOX and the PCBMA shell providing protection. DOX's loading efficiency and content showed consistent patterns with increasing drug concentration. In acidic conditions, DOX was slowly released from the micellar core.

7.4 SMART POLYMER DDS

Smart DDS refers to using stimuli-responsive polymers to change their properties in response to external stimuli. Smart DDS can be created by grafting these polymers onto nanomaterials like silica nanoparticles. Shah et al. [41] investigate different stimuli that can induce the responsiveness of these polymers, including pH, temperature, redox reactions, reactive oxygen species, glucose levels, enzymes, and magnetic fields. The publication provides valuable insights into the structures of diverse polymers that respond to stimuli, a crucial aspect in the advancement of intelligent DDSs.

Szewczyk-Łagodzińska et al. [42] explore atom transfer radical polymerization advancements for synthesizing nanostructured polymeric materials. The authors primarily concentrate on linear and branched block copolymers and bioconjugates, specifically emphasizing intelligent DDSs capable of releasing bioactive substances triggered by external stimuli. These stimuli encompass physical factors like light, ultrasound, or temperature, as well as chemical factors like variations in pH or the environmental redox potential.

Li et al. [43] present a promising therapeutic approach for preventing and treating post-traumatic peritendinous adhesion (PA). They utilize a polylactic acid (PLA)-Dicumarol conjugate-electrospun nanofiber membrane, which achieves spatial accuracy and sustained drug release over time. The membrane acts as a physical barrier while continuously releasing Dicumarol, which inhibits the production of the fibrogenic cytokine transforming growth factor beta (TGFβ) and modulates the fibroblastic TGFβ/Smad3 signaling pathway. This approach reduces tissue adhesion, decreases fibroblast activity, and improves the biomechanical properties of the tendon. Overall, this strategy holds promise as a therapeutic option for individuals suffering from PA.

Ma et al. [44] discuss the smart DDS based on stimuli-responsive polymeric hydrogels. To enhance their mechanical properties, these hydrogels were created using reversible addition-fragmentation chain-transfer polymerization (RAFT) and dual crosslinking methods. Two thermo-sensitive polymers, PH and PG, based on PNIPAM, were synthesized by incorporating β-Cyclodextrin and adamantane groups. The hydrogels were constructed through host-guest interactions and dynamic acyl hydrazone bonding, utilizing commercially available adipic dihydrazide. The resulting hydrogels exhibited sol-gel transitions controlled by pH, temperature-dependent swelling behavior, good stretchability, and self-healing properties.

Aslam et al. [45] developed an intelligent hydrogel capable of controlled release of Acyclovir sodium. The hydrogel was formulated using quince/mucin co-poly(methacrylate) through free-radical polymerization. It exhibited responsive swelling behavior to pH and electrolyte conditions and demonstrated compatibility with the drug based on compatibility studies. The hydrogel enabled sustained drug release for 36 hours, following

Hixen-Crowell's kinetics with a diffusion mechanism governed by Fickian principles.

In a separate study, Zhu et al. [46] designed and synthesized a composite hydrogel of photo-responsive silica polyvinylpyrrolidone (SiO_2-PVP) composite. This composite hydrogel combined the advantageous properties of the SiO_2 framework with the excellent deformability of PVP. Under optimized conditions, the hydrogel demonstrated an elongation capacity of up to 160% and a swelling capability of up to 600%. It exhibited high drug loading capacity and sustained delivery of Ibuprofen for 10 hours in a solution. The article suggests that the hydrogel morphological properties make it suitable for potential applications as a skin patch.

López-Saucedo et al. [47] highlight that modifying polypropylene (PP) film is synthesized as a suitable surface for carrying glycopeptide antibiotics using gamma-irradiation. The process involved exposing the PP to gamma rays and grafting with methyl methacrylate and N-vinyl imidazole through simultaneous irradiation. The modified material was loaded with Vancomycin and tested for release at different pH levels using UV-vis spectrophotometry. The study shows that the modified material has pH-responsive and moderate hydrophilic properties suitable for CDD.

Huang et al. [48] discuss the prevalence of diabetes worldwide and the challenges associated with its treatment. It highlights the potential of phenylboronic acid-based smart DDS in mimicking the functioning of human islets releasing insulin in response to blood glucose levels.

7.5 3D/4D PRINTING

3D printing is the technique of creating a three-dimensional object layer by layer. It utilizes computer-aided design (CAD) software to design a digital model of the object, which is then printed using a 3D printer. This technology has gained significant popularity in biomaterials, specifically for fabricating complex structures such as tissue engineering scaffolds and medical implants.

Saravanou et al. [49] have developed a bio-ink with a dual crosslinked network. This bio-ink is based on a graft copolymer of Sodium alginate and Poly(N-isopropylacrylamide-co-N-tert-butylacrylamide) P(NIPAM-co-NtBAM) side chains. The copolymer undergoes a two-step gelation process, which involves reinforced hydrophobic crosslinking. As a result, the storage modulus is significantly improved, with a five-to-eight-fold increase. The bio-ink can form complex shapes using mild 3D printing conditions.

Additionally, it supports the growth of human periosteum-derived cells in 3D spheroids. The thermal crosslinking of the polymer network can be reversed, facilitating the recovery of the cell spheroids. This innovative bio-ink shows potential for use as a template for forming cell spheroids in 3D fabrication applications.

Chopra et al. [50] focus on Chitosan and its derivatives, which have been investigated as active carriers for drug release. The response of Chitosan-based hydrogels to various stimuli, swelling studies, and drug uptake and diffusion out of the matrices are discussed. The drug uptake by Chitosan-based hydrogels is excellent, and diffusion occurred in the given stimuli with a very high release percentage. The study also explores the applications of 3D printing in drug delivery.

Politakos et al. [51] explore the advancement of 3D printing technology and its potential in 4D printing, which involves using materials capable of responding to environmental stimuli to perform various functions. The application of 3D/4D printing has seen significant growth in areas such as DD, tissue engineering, synthetic organs, and the printing of cells. Polymeric materials are commonly used in these applications, and block copolymers have emerged as a favorable choice due to their ability to combine the desirable properties of each block. This combination improves mechanics, processability, biocompatibility, and responsiveness to stimuli through customizable structures.

7.6 GRAFT COPOLYMERS IN TISSUE REGENERATION

Modifications to the molecular structure of hydrogels allow for customization of their mechanical properties, degradation, mechanism, and cell compatibility, thereby enhancing their potential for tissue regeneration. The latest developments in biopolymer-based hydrogels, especially those with high functionality and integrated DDSs, present notable benefits for clinical applications in tissue regeneration [52].

Castañeda-Rodríguez et al. [53] examine the utilization of poly(lactic acid) (PLA) as a biodegradable polymer. The article addresses the limitations associated with PLA, such as its biological inertness, low cell adhesion, and slow degradation rate. It explores different approaches to modify PLA to overcome these challenges. The authors discuss the development of environmentally friendly biomaterials using renewable resources and the combination of PLA with other polymers as strategies to create new biomaterials with desirable physicochemical properties for tissue engineering applications.

Sedighi et al. [54] investigate the potential applications of glycopolypeptide biopolymers in various biomedical fields, specifically focusing on DDSs. The research emphasizes the self-assembly properties, stable 3D structures, efficient drug loading, and the remarkable biocompatibility and biodegradability of glycopolypeptides. The authors also address the challenges associated with clinical translation and discuss the potential of glycopolypeptide-based nanocarriers in diagnosis and regenerative medicine.

Lu et al. [55] examine the process of cell microencapsulation, which offers cell protection and facilitates nutrient and oxygen exchange while serving as a barrier against the host immune system. Hydrogel encapsulation creates a 3D atmosphere that supports cellular functions and enables tissue structure formation that resembles those in the body. Moreover, the embedded cells can be genetically modified to release targeted therapeutic substances directly at the desired site, minimizing the systemic side effects.

Ilayaperumal et al. [56] discuss using biodegradable polyphosphazene (PPZs) in tissue engineering and biomonitoring. They highlight the versatility of PPZs in synthesis, allowing for the creation of polymers with diverse properties. PPZs have been used as scaffolds for skeletal and tissue regeneration and as carriers for biopharmaceuticals, including anticancer drugs. Incorporating fluorophores into PPZs has expanded their application in tracking tissue regeneration and monitoring the fate of PPZ-based scaffolds and DDSs.

Arun et al. [57] investigate the concept of polymers and their modification through processes such as grafting, bridging, mixing, or the formation of composites. Natural polymers have diverse applications in the culinary and medicinal fields. Polymeric materials are desirable due to their advantageous properties, low density, and capacity to imbibe properties from their components. Chemical bonds within the polymeric structures can be broken by subjecting these materials to high-energy accelerated electrons, forming macromolecule radicals and graft copolymerization. Polymer grafting plays a significant role in the material formulation as it allows for the addition or removal of desired or undesired properties by modifying the functional groups of the polymer.

7.7 OPHTHALMIC NANO DRUG DELIVERY

Graft copolymers are extensively employed in drug delivery applications due to enhanced solubility, stability, and biocompatibility. In the realm of ophthalmic drug delivery, using graft copolymers can improve the bioavailability and effectiveness of drugs by prolonging their retention in the eye.

Liu et al. [58] focus on incorporating immunosuppressants into ophthalmic nano-DDSs for treating autoimmune eye diseases. They highlight the advantages of various delivery systems, including microparticles, nanoparticles, liposomes, micelles, implants, and *in situ* gels. These systems can improve drug solubility, bioavailability, and target specificity while minimizing side effects. The research conducted in this study offers a thorough examination of recent advancements and practical implementations in this field, intending to guide the development of efficient ophthalmic DDS.

Wang et al. [59] discuss the limitations of conventional eye drops in treating ocular diseases due to their lower bioavailability. The authors highlight the benefits associated with nanotechnology, such as an extended duration of drug presence on the surface of the eye, increased penetration through ocular barriers, and enhanced effectiveness of treatments. The article examines typical eye conditions treated using eye drops, the difficulties posed by physiological and anatomical factors during drug delivery, and recent progress in ocular DDSs based on nanotechnology.

Lavikainen et al. [60] modified Gellan gum (GG) by attaching poly(2-ethyl-2-oxazoline) (PEtOx) chains, aiming to create polymeric carriers for ocular drug delivery with reduced mucosal tissue adhesion. The graft copolymers were synthesized using cationic ring-opening polymerization, where living chains were attached to a deprotonated gellan backbone. Varying degrees of grafting were achieved by adjusting the PEtOx grafts to the GG ratio. The successful attachment of PEtOx chains onto the GG backbone was confirmed through spectroscopic and analytical techniques. The obtained copolymer showed a greater diffusion coefficient than the initial GG, indicating potential partial degradation of the polymer backbone or the formation of a more condensed copolymer structure.

7.8 GRAFTING COPOLYMER IN MINIMALLY INVASIVE DRUG DELIVERY

Minimally invasive drug delivery refers to techniques that allow the administration of therapeutic agents with minimal disruption to the body, reducing trauma, pain, and recovery time compared to traditional invasive procedures.

Gheysoori et al. [61] focus on developing thermoresponsive nanocomposite hydrogels for minimally invasive DDS. The study combines Gelatin and Poly(N-isopropylacrylamide) (PNIPAM) to create these hydrogels. PNIPAM nanoparticles are synthesized through photo polymerization and then incorporated into a Gelatin matrix to produce various hybrid nanocomposite hydrogels. The hydrogels exhibit a sol-gel transition at physiological conditions, with reversible behavior around 37°C. By modifying the proportion of PNIPAM and Gelatin, the hydrogels demonstrate changes in the quantity, consistency, and temporal pattern of drug release, indicating their potential for precise drug delivery applications.

7.8.1 TRANSCATHETER ARTERIAL EMBOLIZATION: MINIMALLY INVASIVE TREATMENT

It is a minimally invasive procedure that blocks blood vessels to treat vascular diseases and tumors. Hydrogel-based embolic agents have gained interest due to their potential to overcome limitations. Li et al. [62] provide an advancement in polymer-based hydrogels for endovascular embolization. The discussion covers *in situ* gelling hydrogels, imageable hydrogels, drug-loaded hydrogels, hemostatic hydrogels, shape memory hydrogels, and functional hydrogels. The article provides a comprehensive exploration of these approaches using hydrogels for embolization.

7.9 TRANSDERMAL DRUG DELIVERY (TDD)

Graft copolymers have shown promise in TDD, which involves delivering drugs through the skin for systemic or localized therapeutic effects. TDD offers several advantages, including non-invasiveness, sustained drug release, improved patient compliance, and avoidance of first-pass metabolism.

Qu et al. [63] discuss the global burden of skin diseases and their impact on the global economy. Traditional treatment methods for these conditions often have limitations in efficacy, long treatment durations leading to recurring episodes, and potential systemic side effects due to inadequate drug delivery. However, advanced TDD strategies based on nanocarriers and microneedles offer efficient and targeted drug delivery to the affected skin areas. These strategies minimize systemic toxicity, increase drug concentration in localized skin lesions, and are characterized by non-invasiveness or minimal invasiveness. They offer advantages such as enhanced skin penetration, CDD, targeted delivery, and theranostic functions. The article discusses the design principles, benefits, and considerations of nanocarrier and microneedle-based strategies for TDD. It finds applications in treating different skin ailments.

Sabbagh et al. [64] discuss TDD systems as highly desirable methods for novel drug administration. TDD provides benefits over alternative delivery methods by reducing the risk of liver dysfunction and irritation in the gastrointestinal (GI) tract as potential side effects. It also maintains a consistent and effective drug delivery rate over time. Although various patches, including medical plasters, have been approved worldwide, they have advantages and disadvantages. The study emphasizes the significance of effective design, materials, manufacturing processes, and evaluation methods in utilizing this advanced technology for administering a wide range of drugs.

Li et al. [65] discuss the development of a microneedle (MNs) patch that utilizes a physically entangled crosslinked hydrogel for precise drug release. The hydrogel matrix consists of Gelatin modified with carboxylic end-capped Poly(N-isopropylacrylamide) (PNIPAm), which undergoes a reversible change from a linear to a coiled conformation in response to temperature variations. By utilizing a fast separation method in conjunction with the thermosensitive hydrogel, effective administration of drug-loaded MNs into the skin and controlled drug delivery within the skin can be accomplished in just a few seconds. Insulin was used as a test drug in this system, demonstrating a hypoglycemic effect in diabetic mice through the controlled release of insulin from the patch. The research underscores the viability of this system for the painless and non-invasive delivery of small hydrophilic molecules and peptide medications while minimizing safety concerns.

7.10 GRAFT COPOLYMER APPLICATIONS IN WOUND HEALING

Graft copolymers have been explored for their potential applications in wound healing, as they possess unique properties such as improved mechanical properties, enhanced biocompatibility, and controlled drug delivery capabilities.

In their study, Jha et al. [66] aimed to develop graft copolymers of Polyacrylamide (PAm) with tamarind seed polysaccharide (TSP) using a

32-factorial design. The grafting efficiency, conversion, and initiator effect of the resulting TSP-g-PAm copolymers were evaluated and characterized using various methods. The findings indicated that TSP-g-PAm exhibited enhanced thermal stability and swelling capacity compared to TSP alone and demonstrated antimicrobial activity. The study suggests that TSP-g-PAm has promising potential for advanced eco-friendly materials in applications such as DDS, wound dressing, and the food and cosmetic industries.

In another study, Nezami et al. [67] synthesized a hydrogel by grafting Itaconic acid onto starch in the presence of Fe_3O_4 nanoparticles. The resulting material was thoroughly characterized using various techniques. The incorporation of Fe_3O_4 nanoparticles endowed magnetic properties to the hydrogel, enabling the controlled and pH-dependent binding and release of Guaifenesin (GFN). Additionally, applying an external magnetic field increased the percentage of drug release. The released GFN maintained its bioavailability, suggesting that the hydrogel drug carriers have the potential for magnetically targeted drug delivery and wound healing purposes.

Verma et al. [68] draw attention to the potential of Tragacanth gum, a novel polysaccharide, in developing hydrogels with desirable properties for wound care systems and infection-resistant materials. They emphasize the significance of controlling the physicochemical properties of the grafted matrix by adjusting grafting parameters, enabling the controlled design of advanced materials with various potential applications.

Sharma et al. [69] provide an overview of keratin-based formulations (KFs), including their extraction methods, structure, and characteristics of keratin, which is a bio-sourced polymer found in feathers, hair, hooves, horns, and wool. The article also explores recent advancements in developing KFs, such as 3-D scaffolds, films, fibers, and hydrogels. The potential of blending keratin with other natural and synthetic polymers is discussed as a promising approach to creating innovative keratinous formulations.

7.11 GRAFT COPOLYMER IN INJECTABLE HYDROGELS

Injectable hydrogels, capable of forming a three-dimensional network and retaining water, have garnered significant attention in graft copolymers. These copolymers have diverse applications in tissue engineering, DDS, wound healing, and regenerative medicine. Providing a scaffold-like structure facilitates the minimally invasive administration of therapeutic agents.

Phogat et al. [70] discuss injectable hydrogel scaffolds for efficient bone tissue healing and regeneration. They explore the advantages of injectable scaffolds and the growing trend of developing nanocomposite scaffolds by combining biopolymers and inorganic nanofillers to mimic natural bone's hierarchical nanostructure and composition. The article presents different strategies for creating injectable composites, focusing on nanocomposite hydrogels that incorporate bioinert and bioactive nanofillers. The authors also address the fundamental challenges of developing these injectable hydrogel nanocomposite scaffolds and discuss current research efforts to overcome these obstacles. The article concludes by discussing future trends and providing insights into the development of next-generation injectable hydrogel nanocomposite bone scaffolds.

7.11.1 THERMO-RESPONSIVE INJECTABLE HYDROGELS SELF-ASSEMBLY

Safakas et al. [71] conducted a study on developing injectable hydrogels with thermoresponsive properties. They used graft copolymers based on polysaccharides exhibiting lower critical solution temperatures (LCSTs). By incorporating an alginate-based thermoresponsive gelator with two grafting chains, they controlled the critical gelation temperature of the hydrogel. Adjusting the hydrophobic content of the grafts resulted in a blended side chain structure within the crosslinked nanodomains of the formed network. These unique characteristics make it suitable for biomedical applications, with promising characteristics like injectability and self-healing behavior.

7.12 INJECTABLE HYDROGEL FOR DDS

Mahanta et al. [72] synthesized a hydrogel by modifying chitosan, a highly hydrophilic material, through grafting ester-diol-based polyurethane. The hydrogel formed by the graft copolymer exhibits an interconnected porous network, making it well-suited for drug delivery applications. *In vitro* studies on DDS demonstrated sustained release compared to pure chitosan, with the release kinetics of two model drugs tested. The hydrogel showed biocompatibility and enhanced cellular uptake of drugs, leading to improved efficacy in killing cells. The gelation characteristics of the graft copolymer were validated through experiments using animal models, affirming its suitability as an injectable hydrogel for disease management. The study concludes that the developed graft copolymer can be utilized as an injectable hydrogel for CDD.

A copolymer hydrogel with light and thermo-sensitivity was synthesized by Pourbadiei et al. [2]. The hydrogel was prepared by incorporating an azobenzene derivative and N-isopropyl acrylamide (NIPAM) through host-guest interactions with Cyclodextrin. Modified starch will adjust the hydrogel's LCST, allowing easy injection at low temperatures and gel formation upon local injection. The hydrogel's drug release behavior was studied, and a Paclitaxel-loaded hydrogel was tested against cancerous fibroblastic cells (A-431) *in vitro*. Under 365 nm light irradiation, the hydrogel significantly reduced cell viability. *In vivo* effects were evaluated using histological staining techniques, highlighting the potential of these stimuli-responsive hydrogels for localized drug administration at tumor sites with sustained release properties.

Gao et al. [74] investigate Icaritin (ICT), a natural compound from Epimedium, for treating osteoporosis. ICT exhibits various therapeutic effects, including osteoporosis prevention, bone resorption inhibition, and osteogenic differentiation stimulation. But its poor solubility and PK properties limit its efficacy. The article discusses ICT's osteogenic effects, PKs, and delivery systems, focusing on bone-targeting strategies. The study suggests ICT's potential as a therapeutic agent for osteoporosis.

Parameswaran et al. [75] present a study on synthesizing an injectable thermoresponsive copolymer hydrogel using graft polymerization. The specific copolymer used is hydroxypropyl guar-graft-poly(N-vinylcaprolactam) (HPG-g-PNVCL), which was further modified with nano-hydroxyapatite (n-HA) by a cross-linker called divinyl sulfone. The grafted copolymer and composites obtained were thoroughly characterized using various spectroscopic and analytical techniques. The rheological analysis confirmed that the hydrogel exhibited a low critical solution temperature and thermoreversibility. The hydrogel was utilized for the controlled delivery of the antibiotic drug Ciprofloxacin. Additionally, the hydrogel demonstrated biocompatibility for the growth of osteoblastic cells and the ability to form an apatite-like structure through an *in vitro* biomineralization study.

Koga et al. [76] developed hybrid graft copolymers based on peptides forming injectable hydrogels. These copolymers consisted of a main chain made of poly(2-methacryloyloxyethyl phosphorylcholine) grafted with self-assembling oligoalanine-block-PEG. The copolymers exhibited stable β-sheet structures and had unique properties, including shear-thinning and self-healing abilities. The mechanical properties of the copolymers varied depending on the density of the grafts. These injectable hydrogels were employed to fabricate three-dimensional cell scaffolds and CDD.

Alopaeus et al. [77] devised a film for buccal drug delivery that integrates a self-forming micellar drug solubilizer into the film matrix. To enhance its adhesion to the buccal mucosa, a mucoadhesive polymer was also included in the formulation. The film was fabricated using the solvent casting method and consisted of a graft copolymer (Soluplus®) serving as both the solubilizer and film-forming agent. Additionally, hydroxypropyl methylcellulose (HPMC) or modified hydroxypropyl pea starch (Lycoat®) was used as the mucoadhesive polymer. Combining these polymers in the film improved its erosion resistance and mucoadhesive properties compared to films made with a single polymer. The film formed drug-solubilizing micelles within a hydrogel scaffold that adhered well to the buccal mucosa by incorporating the graft copolymer and mucoadhesive polymer. This allowed for extended retention time and safely enhanced drug delivery.

7.13 ORAL CONTROLLED DRUG DELIVERY

In oral drug delivery, graft copolymers have shown the potential to improve drug solubility, stability, mucoadhesive properties, and bioavailability.

Naeem et al. [78] formulated hydrogels that incorporated inclusion complexes of rutin with Hydroxypropyl-β-cyclodextrin for CDD. Rutin, known for its strong antioxidant properties, suffers from restricted solubility and low oral bioavailability. The hydrogels were prepared using Xanthan gum (XG)-based polymers and characterized using various techniques. The outcomes proved that the hydrogels had high porosity, biodegradability, and controlled drug-release properties at pH 1.2 and 7.4. This study suggests that rutin-containing hydrogels could be a promising approach for preventing and treating diseases associated with oxidative stress.

Rani et al. [79] discuss the challenges of oral lipid delivery and the potential of self-emulsifying drug delivery systems (SEDDS) to overcome these challenges. SEDDS are compositions comprising an oily component, surfactants, co-surfactants, agents facilitating emulsification, and co-solvents. Characterizing SEDDS involves evaluating their size, zeta potential, and surface morphology and maintaining supersaturation to prevent drug precipitation. SEDDS has been effectively utilized for delivering various drugs, including those with anti-cancer, anti-viral, anti-bacterial, immunosuppressant, and natural properties. While some SEDDS drug products are already available in the market, there is still untapped potential in exploring their complete capabilities for oral lipid delivery, highlighting their promising prospects.

Kendre et al. [80] improved the solubility and developed a buccal adhesive DDS for the drug bosentan using Soluplus® as the carrier. The study uses a factorial design to optimize the formulation and assess various parameters. The results show a significant increase in solubility and favorable properties of the buccal adhesive tablets. The optimized formulation demonstrates improved release and confirms the components' suitability.

Mehra et al. [81] conducted a study investigating the application of grafted soy protein isolate (SPI) in the pharmaceutical field. They synthesized a polymeric hydrogel, SPI-g-(AA-co-HPBA), through graft copolymerization and analyzed its properties using various techniques. The hydrogel exhibited favorable characteristics such as noncytotoxic behavior, good porosity, increased thermal stability, and pH-responsive behavior. Additionally, the researchers investigated its potential for sustained drug delivery by studying the release of Ciprofloxacin, specifically for treating bacterial infections in the oral cavity. The findings indicated that the SPI-g-(AA-co-HPBA) hydrogel has promising applications as a controlled DDS for such infections.

Verma et al. [82] sought to improve the efficacy of Sumatriptan succinate, a medication used for migraine treatment, by formulating it into a mucoadhesive film. This approach aimed to enable direct delivery into the bloodstream, bypassing the first-pass metabolism. The researchers employed a central composite design methodology to create and optimize multiple formulations. They considered several dependent variables to achieve the desired properties of the formulations, which includes mucoadhesive strength and folding endurance. Among the formulations, the best formulation demonstrated sustained and controlled drug release, with 77.12% released within 24 hours, indicating its potential as a promising treatment option for migraines.

7.14 GRAFT COPOLYMER IN TARGETED DRUG RELEASE

Targeted drug delivery refers to a medical technique or strategy in which drugs or therapeutic agents are specifically delivered to a particular site or location in the body. The objective of TDD is to enhance the concentration of a drug at the desired location while minimizing its impact on non-target areas, resulting in improved effectiveness and reduced side effects. Selecting a specific TDD approach relies on various factors, including the nature of the drug, the specific disease or condition being treated, and the anatomical location of the target site within the body. These approaches hold great promise for improving the effectiveness of therapies while minimizing systemic side effects.

Hu et al. [83] describe developing a polymer containing many amino groups, enhancing the drug DOX loading rate. Additionally, the polymer contains disulfide bonds, which respond to the redox environment in the TME, allowing for targeted drug release. The nanoparticles have a spherical morphology and are of an appropriate size for systemic circulation. They found non-toxic and good cellular uptake of the polymers. *In vivo* experiments demonstrate their effectiveness in inhibiting tumor growth and minimizing the side effects of DOX.

Zhang et al. [84] have developed a TDD system using self-guided photothermal nanocarriers. The nanocarriers are composed of MoS_2 nanodots rearranged with a copolymer (P(OEGA-co-VBA)) and transferrin (Tf) ligands. This enhances the biocompatibility and drug-loading capacity of the MoS_2 nanodots, while the Tf ligands facilitate the uptake of the nanocarriers by HepG2 cells. These nanocarriers remain stable under physiological conditions but rapidly release the drug DOX triggered by acidic pH and GSH. The MoS_2 nanocarriers exhibit improved therapeutic effects due to their targeted uptake, sensitivity to pH and GSH, and localized heating induced by near-infrared radiation. This designed nanocarrier has the potential for application in multimodal biomedical therapy.

The study conducted by Qian et al. [85] investigated the challenges associated with using PEG-based nanomedicines. These challenges include oxidative damage, immune reactions, and restricted bodily elimination. To address these concerns, the authors suggest considering bioinspired zwitterions as a viable alternative to PEG. Zwitterions are distinguished by an equal number of positively and negatively charged ions. Zwitterionic nanovectors possess biomimetic properties and chemical flexibility, making them biocompatible, resistant to fouling, and capable of prolonged circulation in the bloodstream. Based on their chemical structures, zwitterionic nanovectors are classified into linear, hyperbranched, crosslinked, and hybrid nanoparticles. The authors suggest that through innovative zwitterionic engineering, these nano-vectors can be designed to enhance drug delivery and improve efficacy by combining stealth and other biological functionalities with other nano strategies.

Libánská et al. [86] developed polymer conjugates containing dexamethasone (DEX) drug for TDD in chronic inflammatory conditions like rheumatoid arthritis. The polymer-DEX nanomedicine was carefully controlled for its size to enhance circulation in the bloodstream and facilitate renal elimination. For mouse models of chronic arthritis, these nanomedicines demonstrated therapeutic efficacy by reducing joint inflammation and preventing arthritis-related bone damage, even at lower doses. Linking the drug to the polymer improved blood circulation, resulting in CDD and immediate drug activation, specifically in the affected joints.

Gao et al. [74] investigate the therapeutic potential of Icaritin (ICT), a natural compound extracted from Epimedium, for osteoporosis treatment. Icaritin (ICT) demonstrates a range of effects, including inhibiting bone resorption and promoting bone cell differentiation and maturation. These properties make it a potential candidate for the prevention of osteoporosis. However, the limited solubility, high lipophilicity, and unfavorable PKs of ICT pose challenges to its effectiveness. Researchers are therefore investigating new DDS to overcome these limitations. The article focuses on ICT's osteogenic effects, PKs, and delivery systems, emphasizing strategies to target ICT to specific bone sites.

Naito et al. [73] researched the ideal size of macromolecular drugs that can effectively penetrate and distribute within muscle tissues. This presents a significant hurdle in achieving TDD in muscles. A size-adjustable polymeric delivery system was developed by attaching PEGs to the PAsp backbone, resulting in a series of PEG-grafted copolymers ranging from 11 to 32 nm. The research revealed that the accumulation of these copolymers in muscle tissues declined after reaching a size of approximately 15 nm. When tested in a mouse model of Duchenne muscular dystrophy, the copolymers demonstrated a 2–3-fold increase in accumulation within skeletal muscles compared to normal mice. Imaging and analysis techniques confirmed the extravasation and localized accumulation of the copolymers in skeletal muscles, indicating the potential of macromolecular drugs sized between 11–32 nm for muscle-targeted drug delivery.

Verma et al. [29] researched synthesizing a hydrogel that can release Gentamicin in a controlled manner under different pH conditions. They modified the hydrogel by incorporating boron nitride nanosheets to enhance its adsorption rate. The properties of the hydrogel were investigated using various techniques for chemical interaction, crystallinity, morphology, and thermal degradation. The hydrogel demonstrated compatibility with normal human cells and exhibited pH-dependent drug release behavior. Including boron nitride nanosheets improved drug release. The study presents the potential of hydrogels for TDD in the colon through controlled release mechanisms.

7.15 CONCLUSION

Graft copolymers provide numerous benefits, such as adjustable characteristics, controlled release patterns, enhanced stability, biocompatibility, and the ability to target drug delivery. In cancer therapy, graft copolymers can potentially improve the effectiveness and selectivity of anticancer medications by enhancing their solubility, stability, and targeted transport to cancerous cells. Additionally, these copolymers can be engineered to respond to specific stimuli, such as pH, temperature, or enzymatic activity, enabling the development of intelligent DDSs that release medications at specific locations as desired. For wound healing applications, graft copolymers can provide scaffold structures that promote cell adhesion, proliferation, and tissue regeneration. They are engineered to mimic the extracellular matrix, offering mechanical support and controlled release of growth factors or other bioactive molecules to accelerate wound healing processes. In tissue engineering and regenerative medicine, graft copolymers can serve as scaffolds for tissue regeneration. They can provide a three-dimensional structure that supports cell growth, differentiation, and tissue formation. The properties of graft copolymers can be tailored to match specific tissue requirements, such as mechanical strength, biodegradability, and cell adhesion properties.

Furthermore, graft copolymers can form polymeric micelles, self-assembled structures capable of encapsulating hydrophobic drugs. These micelles can enhance the solubility and stability of poorly water-soluble drugs and improve their bioavailability. Additionally, graft copolymers can be functionalized with targeting ligands to achieve selective drug delivery to specific tissues or cells, minimizing off-target effects and improving therapeutic outcomes. Overall, applying graft copolymers in DDSs for cancer treatment, smart drug delivery, wound healing, scaffolds, targeted drug delivery, polymeric micelles, and tissue regeneration represents a promising avenue in biomedical research. Continued advancements in the graft copolymer design and engineering techniques are expected to improve therapeutic efficacy, patient outcomes, and the overall field of drug delivery and regenerative medicine.

KEYWORDS

- **biocompatibility**
- **biodegradability**
- **controlled drug delivery**
- **drug delivery systems**
- **graft copolymers**

REFERENCES

1. Correa, S., Grosskopf, A. K., Lopez Hernandez, H., Chan, D., Yu, A. C., Stapleton, L. M., & Appel, E. A. (2021). Translational applications of hydrogels. *Chemical Reviews, 121*(18), 11385–11457.
2. Pourbadiei, B., Yousefi Adlsadabad, S., Rahbariasr, N., & Pourjavadi, A. (2023). Synthesis and characterization of dual light/temperature-responsive supramolecular injectable hydrogel based on host-guest interaction between azobenzene and starch-grafted β-cyclodextrin: Melanoma therapy with paclitaxel. *Carbohydrate Polymers, 313,* 120667.
3. Bhattacharya, S., Prajapati, B. G., & Singh, S. (2023). A critical review on the dissemination of pH and stimuli-responsive polymeric nanoparticular systems to improve drug delivery in cancer therapy. *Critical Reviews in Oncology/Hematology, 185*, 103961.
4. Aashli, Giridhar Reddy, S., Siva Kumar, B., Prashanthi, K., & Ananda Murthy, H. C. (2023). Fabricating transdermal film formulations of montelukast sodium with improved chemical stability and extended drug release. *Heliyon, 9*(3), e14469.
5. Muir, V. G., & Burdick, J. A. (2020). Chemically modified biopolymers for the formation of biomedical hydrogels. *Chemical Reviews, 121*(18), 10908–10949.
6. Giridhar Reddy, S. (2023). Kinetic studies for releasing hydroxychloroquine sulphate drug (HCQ) *in vitro* in simulated gastric and intestinal fluids from sodium alginate and lignosulphonic acid blends. *Trends in Sciences, 20*(5), 5318.
7. Giridhar Reddy, S. (2022). Effect of crosslinking on controlled drug release of hydroxychloroquine sulphate drug using alginate beads. *Iranian Journal of Materials Science and Engineering, 19*(2), 1–9.
8. Ulbrich, K., Holá, K., Šubr, V., Bakandritsos, A., Tuček, J., & Zbořil, R. (2016). Targeted drug delivery with polymers and magnetic nanoparticles: Covalent and noncovalent approaches, release control, and clinical studies. *Chemical Reviews, 116*(9), 5338–5431.
9. Pacheco, C., Baião, A., Ding, T., Cui, W., & Sarmento, B. (2023). Recent advances in long-acting drug delivery systems for anticancer drugs. *Advanced Drug Delivery Reviews, 194*, 114724.

10. Rani, R., Malik, P., Dhania, S., & Mukherjee, T. K. (2023). Recent advances in mesoporous silica nanoparticle-mediated drug delivery for breast cancer treatment. *Pharmaceutics, 15*(1), 227.
11. Ehsanimehr, S., Najafi Moghadam, P., Dehaen, W., & Shafiei-Irannejad, V. (2021). Synthesis of pH-sensitive nanocarriers based on polyacrylamide grafted nanocrystalline cellulose for targeted drug delivery to folate receptor in breast cancer cells. *European Polymer Journal, 150*, 110398.
12. Kong, X., Qi, Y., Wang, X., Jiang, R., Wang, J., Fang, Y., Gao, J., & Hwang, K. C. (2023). Nanoparticle drug delivery systems and their applications as targeted therapies for triple-negative breast cancer. *Progress in Materials Science, 134*, 101070.
13. Huang, H. J., Huang, S. Y., Wang, T. H., Lin, T. Y., Huang, N. C., Shih, O., Jeng, U. S., Chu, C. Y., & Chiang, W. H. (2023). Clay nanosheets simultaneously intercalated and stabilized by PEGylated chitosan as drug delivery vehicles for cancer chemotherapy. *Carbohydrate Polymers, 302*, 120390.
14. Massoumi, B., Abbasian, M., Jahanban-Esfahlan, R., Motamedi, S., Samadian, H., Rezaei, A., Derakhshankhah, H., Farnudiyan-Habibi, A., & Jaymand, M. (2020). PEGylated hollow pH-responsive polymeric nanocapsules for controlled drug delivery. *Polymer International, 69*, 519–527.
15. Ramezanpour, A., Ansari, L., & Rahimkhoei, V. (2023). Recent advances in carbohydrate-based paclitaxel delivery systems. *Polymer Bulletin. 81*(2), 1–27.
16. Grebinyk, A., Prylutska, S., Grebinyk, S., Ponomarenko, S., Virych, P., Chumachenko, V., Kutsevol, N., Prylutskyy, Y., Ritter, U., & Frohme, M. (2022). Drug delivery with a pH-sensitive star-like dextran-graft polyacrylamide copolymer. *Nanoscale Advances, 4*, 5077–5088.
17. Khabibullin, V. R., Chetyrkina, M. R., Obydennyy, S. I., Maksimov, S. V., Stepanov, G. V., & Shtykov, S. N. (2023). Study on doxorubicin loading on differently functionalized iron oxide nanoparticles: Implications for controlled drug-delivery application. *International Journal of Molecular Sciences, 24*(5), 4480.
18. Zhang, Y., Kim, Y., Lu, Y., Xu, Y., Yu, D. G., & Song, W. (2022). Intelligent poly(l-histidine)-based nano vehicles for controlled drug delivery. *Journal of Controlled Release, 349*, 963–982.
19. Wang, M. Q., Zou, H., Liu, W. B., Liu, N., & Wu, Z. Q. (2022). Bottlebrush polymers based on RAFT and the "C1" polymerization method: Controlled synthesis and application in anticancer drug delivery. *ACS Macro Letters, 11*(2), 179–185.
20. Shukla, A., Singh, A. P., Ray, B., Aswal, V., Kar, A. G., & Maiti, P. (2019). Efficacy of polyurethane graft on cyclodextrin to control drug release for tumor treatment. *Journal of Colloid and Interface Science, 534*, 215–227.
21. Kostka, L., Kotrchová, L., Šubr, V., Libánská, A., Ferreira, C. A., Malátová, I., Lee, H. J., Barnhart, T. E., Engle, J. W., Cai, W., Šírová, M., & Etrych, T. (2020). HPMA-based star polymer biomaterials with tunable structure and biodegradability tailored for advanced drug delivery to solid tumors. *Biomaterials, 235*, 119728.
22. El-Zeiny, H. M., Abukhadra, M. R., Sayed, O. M., Osman, A. H. M., & Ahmed, S. A. (2020). Insight into novel β-cyclodextrin-grafted-poly (N-vinylcaprolactam) nanogel structures as advanced carriers for 5-fluorouracil: Equilibrium behavior and pharmacokinetic modeling. *Colloids and Surfaces A: Physicochemical and Engineering Aspects, 586*, 124197.

23. Bakr, E. A., Gaber, M., Saad, D. R., & Salahuddin, N. (2023). Comparative study between two different morphological structures based on polylactic acid, nanocellulose, and magnetite for co-delivery of fluorouracil and curcumin. *International Journal of Biological Macromolecules, 230*, 123315.
24. Demirel, E., & Yuksel Durmaz, Y. (2023). PEGylated reduced graphene oxide as nanoplatform for targeted gene and drug delivery. *European Polymer Journal, 186*, 111841.
25. Bláhová, M., Randárová, E., Konefał, R., Nottelet, B., & Etrych, T. (2020). Graft copolymers with tunable amphiphilicity tailored for efficient dual drug delivery via encapsulation and pH-sensitive drug conjugation. *Polymer Chemistry, 11*(21), 4438–4453.
26. Qiu, C., Wu, Y., Shi, Q., Guo, Q., Zhang, J., Meng, Y., Wang, C., Xia, F., Wang, J., & Xu, C. (2023). Advanced strategies for nucleic acids and small-molecular drugs in combined anticancer therapy. *International Journal of Biological Sciences, 19*(3), 789–810.
27. Tao, W., Wang, J., Zhou, Y., Liu, Z., Chen, H., Zhao, Z., Yan, H., Liao, X., & Zhang, L. (2023). Acid/reduction dual-sensitive amphiphilic graft polyurethane with folic acid and detachable poly (ethylene glycol) as anticancer drug delivery carrier. *Colloids and Surfaces B: Bio Interfaces, 222*, 113084.
28. Movagharnezhad, N., Ehsanimehr, S., & Najafi Moghadam, P. (2022). Synthesis of poly (N-vinylpyrrolidone)-grafted-magnetite Bromo acetylated cellulose via ATRP for drug delivery. *Materials Chemistry Horizons, 1*(2), 89–98.
29. Verma, A., Sharma, B., Kalia, S., Alsanie, W. F., Thakur, S., & Thakur, V. K. (2023). Carboxymethyl cellulose-based sustainable hydrogel for colon-specific delivery of gentamicin. *International Journal of Biological Macromolecules, 228*, 773–782.
30. Monajati, M., Tamaddon, A. M., Abolmaali, S. S., Yousefi, G., Borandeh, S., & Dinarvand, R. (2023). Enhanced L-asparaginase stability through immobilization in supramolecular nano gels of PEG-grafted poly HPMA with bis(α-cyclodextrin). *Biochemical Engineering Journal, 191*, 108802.
31. Kashyap, M. (2023). Advances in polymer optimization for enhanced drug delivery. In S. K. Singh, M. Gulati, S. Mutalik, M. Dhanasekaran, & K. Dua (Eds.), *Polymeric Micelles: Principles, Perspectives and Practices* (pp. 69–92). Springer.
32. Mahajan, D., Suares, D., Rachana, R., & Shetty, S. (2023). Types of polymeric micelles for controlled drug release. In S. K. Singh, M. Gulati, S. Mutalik, M. Dhanasekaran, & K. Dua (Eds.), *Polymeric Micelles: Principles, Perspectives and Practices* (pp. 93–111). Springer.
33. Hari, S. K., Gauba, A., & Shrivastava, N. (2023). Polymeric micelles and cancer therapy: An ingenious multimodal tumor-targeted drug delivery system. *Drug Delivery and Translational Research, 13*, 135–163.
34. Mehan, N., Kumar, M., Bhatt, S., & Shankar, R. (2022). Self-assembly polymeric nano micelles for the futuristic treatment of skin cancer and phototoxicity: Therapeutic and clinical advancement. *Therapeutic Drug Carrier Systems, 39*(2), 79–95.
35. Adeli, F., Abbasi, F., & Babazadeh, M. (2022). Thermo/pH dual-responsive micelles based on the host–Guest interaction between benzimidazole-terminated graft copolymer and β-cyclodextrin-functionalized star block copolymer for smart drug delivery. *Journal of Nanobiotechnology, 20*(1), 91.
36. Unde, J. S., & Shukla, R. (2023). Polymeric micelles in the delivery of therapeutic phytoconstituents. In S. K. Singh, M. Gulati, S. Mutalik, M. Dhanasekaran, & K. Dua

(Eds.), *Polymeric Micelles: Principles, Perspectives and Practices* (pp. 169–183). Springer.

37. Niesyto, K., Mazur, A., & Neugebauer, D. (2022). Dual-drug delivery via the self-assembled conjugates of choline-functionalized graft copolymers. *Materials, 15*(13), 4457.
38. Kuperkar, K., Patel, D., Atanase, L. I., & Bahadur, P. (2022). Amphiphilic block copolymers: Their structures and self-assembly to polymeric micelles and polymersomes as drug delivery vehicles. *Polymers, 14*(21), 4702.
39. Singam, A., Killi, N., Patel, P. R., & Gundloori, R. V. N. (2021). PEGylated ethyl cellulose micelles as a nanocarrier for drug delivery. *RSC Advances, 11*, 30532–30543.
40. Zeng, S., Quan, X., Zhu, H., Sun, D., Miao, Z., Zhang, L., & Zhou, J. (2021). Computer simulations on a pH-responsive anticancer drug delivery system using zwitterion-grafted polyamidoamine dendrimer unimolecular micelles. *Langmuir, 37*(3), 1225–1234.
41. Shah, I. U., Jadhav, S. A., Belekar, V. M., & Patil, P. S. (2023). Smart polymer grafted silica-based drug delivery systems. *Polymer Advanced Technologies, 34*(1), 24–43.
42. Szewczyk-Łagodzińska, M., Plichta, A., Dębowski, M., Kowalczyk, S., Iuliano, A., & Florjańczyk, Z. (2023). Recent advances in the application of ATRP in the synthesis of drug delivery systems. *Polymers, 15*(5), 1234.
43. Li, Y., Hu, C., Hu, B., Tian, J., Zhao, G., Cai, C., Sun, Z., Wang, S., Pang, S., Bao, R., Tao, Z., Chen, H., Wu, J., & Liu, S. (2023). Sustained release of dicumarol via novel grafted polymer in electrospun nanofiber membrane for treatment of peritendinous adhesion. *Advanced Healthcare Materials, 12(15)*, e2203078.
44. Ma, W., Yang, X., Liu, H. B., Guo, Z. R., Zhang, J. L., Kang, G. Y., Yu, C. Y., & Wei, H. (2023). Fabrication of thermo and pH-dual sensitive hydrogels with optimized physicochemical properties via host-guest interactions and acyl hydrazone dynamic bonding. *Reactive and Functional Polymers, 184*, 105513.
45. Aslam, A., Ashraf, M. U., Barkat, K., Mahmood, A., Hussain, M. A., Farid-ul-Haq, M., Lashkar, M. O., & Gad, H. A. (2023). Fabrication of stimuli-responsive quince/mucin co-poly(methacrylate) hydrogel matrices for the controlled delivery of acyclovir sodium: Design, characterization, and toxicity evaluation. *Pharmaceutics, 15*(2), 650.
46. Zhu, W., Zhao, Y., Wu, Z., Lv, F., Zhang, Y., & Guo, S. (2023). Application of UV responsive SiO_2/PVP composite hydrogels as intelligent controlled drug release patches. *Polymer, 264*, 125535.
47. López-Saucedo, F., López-Barriguete, J. E., Flores-Rojas, G. G., Gómez-Dorantes, S., & Bucio, E. (2022). Polypropylene graft poly (methyl methacrylate) graft poly(N-vinylimidazole) as a smart material for pH-controlled drug delivery. *International Journal of Molecular Sciences, 23*(1), 304.
48. Huang, Q., Wang, L., Yu, H., & Ur-Rahman, K. (2019). Advances in phenylboronic acid-based closed-loop smart drug delivery system for diabetic therapy. *Journal of Controlled Release, 305*, 50–64.
49. Saravanou, S. F., Ioannidis, K., Dimopoulos, A., Paxinou, A., Kounelaki, F., Varsami, S. M., Tsitsilianis, C., Papantoniou, I., & Pasparakis, G. (2023). Dually crosslinked injectable alginate-based graft copolymer thermoresponsive hydrogels as 3D printing bioinks for cell spheroid growth and release. *Carbohydrate Polymers, 312*, 120790.
50. Chopra, L., & Chohan, J. S. (2023). Additive manufacturing processes: Role of 3D printing and chitosan-hydrogel-based modifiers in drug delivery. *Additive Manufacturing.*

51. Politakos, N. (2023). Block copolymers in 3D/4D printing: Advances and applications as biomaterials. *Polymers, 15*(2), 322.
52. Hama, R., Ulziibayar, A., Reinhardt, J. W., Watanabe, T., Kelly, J., & Shinoka, T. (2023). Recent developments in biopolymer-based hydrogels for tissue engineering applications. *Biomolecules, 13*(2), 280.
53. Castañeda-Rodríguez, S., González-Torres, M., & Ribas-Aparicio, R. M. (2023). Recent advances in modified poly (lactic acid) as tissue engineering materials. *Journal of Biological Engineering, 17*(1), 21.
54. Sedighi, M., Mahmoudi, Z., Ghasempour, A., Shakibaie, M., Ghasemi, F., Akbari, M., Abbaszadeh, S., Mostafavi, E., Santos, H. A., & Shahbazi, M. A. (2023). Nanostructured multifunctional stimuli-responsive glycopolypeptide-based copolymers for biomedical applications. *Journal of Controlled Release, 354*, 128–145.
55. Lu, T., Xia, B., & Chen, G. (2023). Advances in polymer-based cell encapsulation and its applications in tissue repair. *Biotechnology Progress*, e3325.
56. Ilayaperumal, P., Chelladurai, P., Vairan, K., Anilkumar, P., & Balagurusamy, B. (2023). Polyphosphazenes—A promising candidate for drug delivery, bioimaging, and tissue engineering: A review. *Macromolecular Materials & Engineering, 308(4), 2200553.*
57. Singh, A. K., & Malviya, R. (2023). Graft copolymers of polysaccharide: Synthesis methodology and biomedical applications in tissue engineering. *Current Pharmaceutical Biotechnology, 24*(4), 510–531.
58. Liu, Y., Xu, H., Yan, N., Tang, Z., & Wang, Q. (2023). Research progress of ophthalmic preparations of immune suppressants. *Drug Delivery, 30*(1), 30–48.
59. Wang, C., & Pang, Y. (2023). Nano-based eye drops: Topical and non-invasive therapy for ocular diseases. *Advanced Drug Delivery Reviews, 194*, 114721.
60. Lavikainen, J., Dauletbekova, M., & Toleutay, G. (2021). Poly(2-ethyl-2-oxazoline) grafted gellan gum for potential application in transmucosal drug delivery. *Polymer Advanced Technologies, 32*, 2770–2780.
61. Gheysoori, A., Paydayesh, A., Jafari, M., & Peidayesh, H. (2023). Thermoresponsive nanocomposite hydrogels based on gelatin/poly(N–isopropylacrylamide) (PNIPAM) for controlled drug delivery. *European Polymer Journal, 186*, 111846.
62. Li, X., Ullah, M. W., Li, B., & Chen, H. (2023). Recent progress in advanced hydrogel-based embolic agents: From rational design strategies to improved endovascular embolization. *Advanced Healthcare Materials, 12(7)*, e2202787.
63. Qu, F., Geng, R., Liu, Y., & Zhu, J. (2022). Advanced nanocarrier- and microneedle-based transdermal drug delivery strategies for skin diseases treatment. *Theranostics, 12*(7), 3372–3406.
64. Sabbagh, F., & Kim, B. S. (2022). Recent advances in polymeric transdermal drug delivery systems. *Journal of Controlled Release, 341*, 132–146.
65. Li, J. Y., Feng, Y. H., He, Y. T., Hu, L. F., Liang, L., Zhao, Z. Q., Chen, B. Z., & Guo, X. D. (2022). Thermosensitive hydrogel microneedles for controlled transdermal drug delivery. *Acta Biomaterialia, 153*, 308–319.
66. Jha, S., Malviya, R., Fuloria, S., Sundram, S., Subramaniyan, V., Sekar, M., Sharma, P. K., Chakravarthi, S., Wu, Y. S., Mishra, N., Meenakshi, D. U., Bhalla, V., Djearamane, S., & Fuloria, N. K. (2022). Characterization of microwave-controlled polyacrylamide graft copolymer of tamarind seed polysaccharide. *Polymers, 14*(5), 1037.

67. Nezami, S., Sadeghi, M., & Mohajerani, H. (2020). A novel pH-sensitive and magnetic starch-based nanocomposite hydrogel as a controlled drug delivery system for wound healing. *Polymer Degradation and Stability, 179*, 109255.
68. Verma, C., Pathania, D., Anjum, S., & Gupta, B. (2020). Smart designing of tragacanth gum by graft functionalization for advanced materials. *Macromolecular Materials & Engineering, 305*(7), 1900762.
69. Sharma, S., Rostamabadi, H., Gupta, S., Nadda, A. K., Kharazmi, M. S., & Jafari, S. M. (2022). Nano/micro-formulations of keratin in biocomposites, wound healing, and drug delivery systems; recent advances in biomedical applications. *European Polymer Journal, 180*, 111614. https://doi.org/10.1016/j.eurpolymj.2022.111614
70. Phogat, K., Ghosh, S. B., & Bandyopadhyay-Ghosh, S. (2023). Recent advances on injectable nanocomposite hydrogels towards bone tissue rehabilitation. *Journal of Applied Polymer Science, 140*(4), e53362.
71. Safakas, K., Saravanou, S. F., Iatridi, Z., & Tsitsilianis, C. (2023). Thermo-responsive injectable hydrogels formed by self-assembly of alginate-based heterograft copolymers. *Gels, 9*(3), 236.
72. Mahanta, A. K., & Maiti, P. (2019). Injectable hydrogel through hydrophobic grafting on chitosan for controlled drug delivery. *ACS Applied Bio Materials, 2*(12), 5415–5426.
73. Naito, M., Watanuki, Y., Toh, K., Yum, J., Kim, B. S., Taniwaki, K., Ogura, S., Ishida, H., Cho, M., Chaya, H., Miyajima, K., Yamasaki, Y., Osada, K., Minegishi, K., Aoki, Y., & Miyata, K. (2022). Size-tunable PEG-grafted copolymers as a polymeric nano ruler for passive targeting muscle tissues. *Journal of Controlled Release, 347*, 607–614.
74. Gao, L., & Zhang, S. Q. (2022). Antiosteoporosis effects, pharmacokinetics, and drug delivery systems of icaritin: Advances and prospects. *Pharmaceuticals, 15*(4), 397.
75. Parameswaran-Thankam, A., Parnell, C. M., Watanabe, F., RanguMagar, A. B., Chhetri, B. P., Szwedo, P. K., Biris, A. S., & Ghosh, A. (2018). Guar-based injectable thermoresponsive hydrogel as a scaffold for bone cell growth and controlled drug delivery. *ACS Omega, 3*(11), 15158–15167.
76. Koga, T., Matsuoka, T., Morita, Y., & Higashi, N. (2021). Injectable hydrogels self-assembled from oligopeptide-poly(2-methacryloyloxyethyl phosphorylcholine) hybrid graft copolymers for cell scaffolds and controlled release applications. *Materials Advances, 2*, 4068–4074.
77. Alopaeus, J. F., Hellfritzsch, M., Gutowski, T., Scherlie, R., Almeida, A., Sarmento, B., Škalko-Basnet, N., & Tho, I. (2020). Mucoadhesive buccal films based on a graft copolymer–A mucin-retentive hydrogel scaffold. *European Journal of Pharmaceutical Sciences, 142*, 105142.
78. Naeem, A., Yu, C., Zang, Z., Zhu, W., Deng, X., & Guan, Y. (2023). Synthesis and evaluation of rutin–Hydroxypropyl β-cyclodextrin inclusion complexes embedded in xanthan gum-based (HPMC-g-AMPS) hydrogels for oral controlled drug delivery. *Antioxidants, 12*(3), 552.
79. Rani, S., Rana, R., & Saraogi, G. K. (2019). Self-emulsifying oral lipid drug delivery systems: Advances and challenges. *AAPS PharmSciTech, 20*(1), 129.
80. Kendre, P. N., Chaudhari, P. D., & Jain, S. P. (2021). An effort to augment solubility and efficiency of the oral bosentan-bucco-adhesive drug delivery system using graft copolymer as the carrier. *Polymer Bulletin, 78*, 5851–5871.
81. Mehra, S., Nisar, S., Chauhan, S., Singh, V., & Rattan, S. (2020). Soy protein-based hydrogel under microwave-induced grafting of acrylic acid and 4-(4-hydroxyphenyl)

butanoic acid: A potential vehicle for controlled drug delivery in oral cavity bacterial infections. *ACS Omega, 5*(34), 21610–21622.

82. Verma, S., Tonk, R. K., Albratty, M., Alhazmi, H. A., Najmi, A., Kumar, R., Kumar, M., Taleuzzaman, M., Swami, G., & Alam, M. S. (2022). Design and evaluation of sustained release mucoadhesive film of sumatriptan succinate containing grafted copolymer as the platform. *Saudi Pharmaceutical Journal, 30*(11), 1527–1537.
83. Hu, Z., Wang, G., Zhang, R., Wang, L., Wang, J., Hu, J., & Reheman, A. (2023). Construction of poly(amino acid)s nano-delivery system and sustained release with redox-responsive. *Colloids and Surfaces B: Biointerfaces, 224*, 113232.
84. Zhang, A., Li, A., Zhao, W., Yan, G., Liu, B., Liu, M., Li, M., Huo, B., & Liu, J. (2018). An efficient and self-guided chemo-photothermal drug loading system based on copolymer and transferrin decorated MoS_2 nanodots for dually controlled drug release. *Chemical Engineering Journal, 342*, 120–132.
85. Qian, H., Wang, K., Lv, M., Zhao, C., Wang, H., Wen, S., Huang, D., Chen, W., & Zhong, Y. (2022). Recent advances on next generation of polyzwitterion-based nano-vectors for targeted drug delivery. *Journal of Controlled Release, 343*, 492–505.
86. Libánská, A., Randárová, E., Skoroplyas, S., Bartoš, M., Luňáčková, J., Lager, F., Renault, G., Scherman, D., & Etrych, T. (2023). Size-switchable polymer-based nanomedicines in the advanced therapy of rheumatoid arthritis. *Journal of Controlled Release, 353*, 30–41.

Index

C

D

E

I

P

U

V

W

X

Z